John Gamgee

Yellow Fever

A Nautical Disease

John Gamgee

Yellow Fever
A Nautical Disease

ISBN/EAN: 9783337405236

Printed in Europe, USA, Canada, Australia, Japan

Cover: Foto ©berggeist007 / pixelio.de

More available books at **www.hansebooks.com**

YELLOW FEVER

A NAUTICAL DISEASE.

ITS ORIGIN AND PREVENTION.

BY

JOHN GAMGEE.

"As far as we know, low temperature is the only agency that can be relied on safely to destroy the infection of this disease."—*Dr. Carpenter.*
"Frost puts an end suddenly to our epidemics. Art never can do better than to imitate nature."—*Dr. J. C. Faget.*

NEW YORK:
D. APPLETON AND COMPANY,
549 AND 551 BROADWAY.
1879.

MRS. ELIZABETH THOMPSON.

MADAM: Mankind is forgetful. Perhaps I should say mankind is exacting. Every generous thought and every noble deed, assimilated for man's growth and development, is lost to sight in the very act.

> " Time hath a wallet at his back,
> Wherein he puts alms for oblivion ;
> A great siz'd monster of ingratitudes :
> Those scraps are good deeds past ; which are devoured
> As fast as they are made, forgot as soon
> As done."

You caused the investigation of the Yellow Fever Epidemic of 1878. You started our dead friend, John M. Woodworth, in the excellent work which demanded one more martyr. His willing, sleepless brain and frame, stirred by your impulse, found rest only in death.

What remains? For two centuries the gnawing canker of commerce, on the Atlantic coast of America, has been that pestilence, destructive to more property than lives, but which is so well calculated by its appearance, in bustling seaports, to create panics and distract the wisest councilors.

You have brought order out of chaos. A Central Authority now coördinates for good the many willing, who, seeking the best, often attained the worst. The result is manifest. The public will be saved from that never-ending conflict of medical opinion, powerful in crushing bookshelves and powerless to save one human life. Ultimately and soon Yellow Fever will be counted with the plagues that were.

A common object, a common purpose, a common good—aimed at and secured with the same ease as touching a button atomized the Hell Gate rocks. To you the country owes this touch—a touch of human nature making the whole world kin.

With grateful appreciation of your kind friendship, I remain, madam,

Your obedient servant,

JOHN GAMGEE.

WASHINGTON, D. C., *August 18, 1879.*

PREFACE.

A SOUND medical philosophy is fast becoming a reality. Whoever in the future may prove its eloquent exponent, only a John Hunter or a Helmholtz can be equal to the task. It demands a man of true genius, chained to the car of experimental physics, capable of piercing the bubbles of tradition, strong in marshaling facts and figures, prompt and clear with his relentless logic.

The science of thermo-dynamics had to be born to render this possible. Lomonossow, Rumford, Joule, Mayer, Clausius, Rankine, Sir William Thomson, and Hirn had to do their work. This science is no branch of other sciences. It is the ground-work on which astronomy, biology, pathology, and all other exact studies must be based. Writers like Dr. La Roche, whose work on Yellow Fever is a monument of painstaking zeal and industry, will continue to amass data, so slightly clouded by the fashionable opinions of their time that the vigorous reaping machine of the future may garner ready-stacked treasures.

Every reasonable hypothesis must expedite the future demonstrations. It is absolutely essential to broad scientific development. The record of known truths and the views propounded in the following pages may therefore serve an important purpose. With the confidence of just perception, guided by a very varied experience, I commend to all the means whereby the prompt extinction of yellow fever on the Atlantic coast of the New World is, in my humble opinion, a matter of certainty.

To Dr. J. C. Faget, of New Orleans, now pertains the credit of having first formulated the idea of resorting to artificial refrigeration for the purification of ships. His clear exposition of the probable origin and definite course of yellow fever, in the shipping and seaport towns, has recently given precision and directness to my labors. His name and writings were unknown to me a few weeks since. I shall henceforth hold both in grateful remembrance.

RIGGS HOUSE, WASHINGTON, D. C., *August 22, 1879.*

CONTENTS.

8 CONTENTS.

PAGE

CHAPTER II.

GENERAL HISTORICAL CONSIDERATIONS: Past History of Yellow Fever—Ancient
History of Yellow Fever—Diseases of the Red Men—Spanish and Portu-
guese Traders—The Spanish Galleons—Sir Walter Raleigh's Voyage—Anson's Voyage around the World—Ships composing the Squadron—Anson's
Fleet Surgeon's Opinion—Outbreak of the Calenture—What is a Calenture?—Rest at St. Catherine's—Scurvy in the Pacific—Capture of Pizarro's Galleon—No Yellow Fever in the Pacific—Captain Cook's Voyage—
Pringle's Eulogy—Bryan Edwards on West Indian Fever—Yellow Fever in
British and French Guiana—Yellow Fever on the African Coast—Yellow
Fever in Charleston—European Outbreaks of Yellow Fever—Recent Outbreaks in France and England—Clean Bills of Health—Special Data relating to the West India Islands 74

CHAPTER III.

CONTRAST AND COMPARISONS BETWEEN YELLOW FEVER AND OTHER DISEASES:
Relapsing and Yellow Fever—Remittent and Yellow Fever—The Pulse
and Thermometer in Yellow Fever—Dr. J. C. Faget's Observations—Epidemic Fever of Young Children in New Orleans—Oroya Fever—Cholera
and Yellow Fever. Contrast of a Land and an Ocean Plague—Analogy
between Typhus and Yellow Fever in a Ship 113

CHAPTER IV.

DEVELOPMENTAL PHENOMENA: Where and how may Yellow Fever develop?—
Statement of the Germ Theory—Unformed Ferments—Germs unrecogniz-
able—Lower Organisms not specifically injurious, nor structurally characteristic of Fevers 137

CHAPTER V.

THE PREVENTION OF YELLOW FEVER: Making Quarantines pay—Destruction of
Yellow-Fever Germs by Cold—Artificial Refrigeration proposed by Dr. J.
C. Faget—Cold as an Antidote endorsed by the Board of Experts—How is
it proposed to deal with Shipping?—Little Danger from Cargo—Thorough
Disinfection essential—Why not attack the Disease off Land?—The
Classification of Ships—Special Ships for Special Cargo—Is it necessary
to distinguish between Infected and Non-Infected Vessels?—Cold the
Natural Antidote of Yellow Fever—How to produce and apply the Cold—
Disinfection of Air—Disinfection of the Bilge—Disinfection of Cargo—
Refrigerating Machinery—Store of Cold—Cost of Refrigeration—Report
of the Naval Engineers on the Refrigerating Ship—Cooling and ventilating
Ships—High Temperature as a Yellow-Fever Antidote—Fire as a Disinfec-
tant—Ventilation of Ships by an Injector—Prevention of Yellow Fever in
Cities—Cold and Ventilation—International Coöperation . . 150

CHAPTER VI.

CONCLUSIONS 195

YELLOW FEVER.

INTRODUCTION.

A NAUTICAL disease, a product of foul ships in the equatorial Atlantic caldron, yellow fever, the dreaded scourge of American commerce, is far more easily, surely, and permanently preventable than typhoid or scarlatina. The sewage fever of cities may for centuries afflict humanity. With zeal and firmness, coöperation and persistent effort, a decade should practically abolish the ocean pestilence. Mystery where there should be no mystery—doubt where rays of discriminating light can penetrate—a mountain of a molehill—fairly characterize the involved doctrines which pass current as to the genesis and nature of yellow fever.

My temerity and apparent dogmatism, in a sketch fashioned and almost written at one sitting, may startle many. Such a sketch may prove serviceable. It is always instructive to view a question from all points, and I believe that mine is not the least commanding eminence whence to photograph fresh images of this dire disease, since Audouard, Humboldt, Copland, Currie, La Roche, and especially Dr. Faget, of New Orleans, have indicated how much there is of truth in regarding yellow fever as something more than, and something different from, a land disease.

No recondite research is needed to reveal a new elixir of life. Fresh air, dry air, cold air, available in all parts, driven through and through the stagnant pest-holds of vessels, suffice. The details hereafter.

Pure and frigid air must and can rid the mercantile marine of this ship-typhus—the analogue, in a sense, of that jail pestilence which John Howard banished from British prisons.

It is sometimes well in medicine as in politics to sink minor differences, and strike for some reform which can not fail to commend itself to the vast majority. In this spirit I advocate that natural antidote which, during epidemics in this country, has been universally recognized as all-sufficient with approaching winter.

Stumbling into volunteer service on this question, no time has been lost, no thought or trouble spared, since last December. With La Roche I might say that, " painful as the avowal may be, it is a fact, the truth of which can not be denied, that, notwithstanding all that has been written on the subject of the yellow fever in this and other countries—all the labor that has been bestowed, or any investigation of its causes, character, and anatomical phenomena—little progress has so far been made in a knowledge of the pathology of that disease." May not the fact reveal itself that the disease, in its real origin, has been necessarily less investigated, because that origin is at sea, and the most skillful and painstaking observers are on land? A knowledge of the climate and diseases of the sea, in the pestilential mercantile craft of the tropical Atlantic, acquired by an expert in pathological physics of the modern school, would, if in our possession, add a long, brilliant, and marvelously enlightening chapter to the many penned by city physicians.

In chemistry we have to adopt at one time the analytical, at another the synthetical, method in ascertaining the nature of substances. In pathology, touching such a question as yellow fever, investigations of all kinds must be instituted ; but, while we are told so much—and so much that is conflicting—concerning the causes and development of the disease, I for one think it high time that a combined international effort be made to demonstrate, by complete and practical measures, how far perfect ventilation and cleanliness of ships will contribute to exterminate yellow fever. It is my conviction that it will do so as effectually as prison sanitation has made us strangers to the pestilential typhus of foul dungeons. To clear the Atlantic seaboard of such a plague, the medical authority of one country, however eminent it may be, can not assert itself with the desired promptitude, nor without measures which might be resented. But all—and notably the Brazilian Emperor—might be expected promptly to lend intelligent and financial aid to rid the commerce of the New World of its fatal obstruction—the withering, palsying scourge of seaports and of the noblest seacoast under heaven.

The humblest and the mightiest may join in good works; and as a simple interpreter, according to my lights, of some passages of epidemic history, I commend to all the one great practical recommendation, viz.: the instant and absolute purification of the merchant marine, at all hazards, between New York and Rio. It is not light work, but on the other hand it is definite. The remedy in this case can not fail to prove far better and infinitely less costly than the disease.

CHAPTER I.

Every disease afflicting man and animals admits of being grouped under one of three heads. A contagious and commonly malignant class embodies all essentially migratory maladies, communicated by a specific virus from the sick to the healthy. They can be reproduced at will by collecting and inoculating this virus, which is endowed with a latent vitality equal, in some cases and under suitable conditions, to the seeds of plants which have germinated after preservation in the tissues of Egyptian mummies.* These specific diseases are prop-

---

* The diseases of plants supply us with highly instructive examples. In the vine disease due to *Phylloxera vastatrix*, the living organism inducing and propagating it is an insect which undergoes a variety of metamorphoses, and eluded for long the vigilance of naturalists by its varied abodes in different stages of its existence. The native American vines on which the phylloxera lives, without destroying the plants, are an illustration of immunity by "survival of the fittest," which enables scientific men to propose a remedy wherever the malady has been introduced. Professor Riley, of the United States Entomological Commission, having discovered that the cultivated American vines possessed a varying degree of resistance to the disease, recommended that those least susceptible to it should be used in the French vineyards as stocks, on which to graft their own vines. The demand for cuttings of such American vines has this year exceeded the supply. All other remedies are being abandoned in France, and by means of the American vines there is hope of restoring the blighted vineyards. This serious malady was introduced into France by vine-cuttings after the civil war, and it threatens Spain, Portugal, Switzerland, Austria, and Prussia. Its complete history is highly suggestive to the pathologist, as a type of contagion with manifest traveling germ. The wingless phylloxera travels over the surface of the ground from vine to vine, or beneath the ground where roots interlock; while in the winged form it may fly or be carried as many as fifteen or twenty miles, and under exceptional conditions even more. Through

agated through time and space by contagion, and by contagion alone. The leading types are small-pox in all its varied forms, as afflicting men and animals; the lung-plague of cattle; hydrophobia in men and animals; and certain parasitic maladies. These are essentially INTERNATIONAL in their manifestations—may be exterminated anywhere, and, once exterminated in any locality, can only recur by reimportation, recommunication from the sick to the healthy.

A second and much larger group includes some of the most devastating plagues, but the fundamental character in common is localization as to origin. There may be, and there is, some uncertainty as to the precise line of demarkation between these diseases and the pure contagia. It is the same in every classification of animals and plants themselves. But history records periodic manifestations always traced to definite regions. A susceptible man may be inoculated with small-pox anywhere; a bullock will take the lung-plague in any latitude and at any altitude, and there is practically no difference known in the type and severity of the disease wherever it may occur. The localized endemic or enzoötic maladies are influenced materially and radically by season, temperature, humidity, soil, systems of agriculture, sanitary conditions (especially in relation to disposal of sewage and the water-supply), the geographical distribution of animals and plants, and other secondary circumstances too numerous to mention. These maladies travel and sometimes widely; but their main centers of development are known or knowable, and their common method of propagation usually involves the contamination of air, water, food, or soil to breed a pestilence. It is obvious at once that the leading and strikingly characteristic types of this group are the cholera of Hindostan; the ancient plagues of the East, derived from Africa, and passing through Egypt to desolate the Roman Empire; the yellow fever of the tropical Western Atlantic basin; the varied forms of malignant anthrax (due to great heat and humidity, seriously

man's agency, by commerce in plants and cuttings, it may be carried to indefinite distances. It has always been by man's agency that Asiatic cattle-plague, reputed harmless in the Steppe herds, has been communicated westward as far as the British Isles. The general similarity of the diseases of plants and animals indicates that they are susceptible of similar classification.

aggravated by filth and foul air and known to us as the Siberian boil plague), the bubonic fever or pestilence, etc.; the splenic fever of Texan cattle, pervading perennially a wide region of natural grazing ground, beyond which, in the healthiest localities, the excreta of the apparently healthy and certainly thriving animals contaminate the grasses, and destroy, with the fury of genuine plague, the sound kine of northern latitudes; ague; phthisis; and (as a last example), we may adduce certain parasitic maladies—not those dependent on acari which live in any latitude, but the tropical screw-worm invading every sore and festering, to bodily death, every simple wound. Think of the South African tsetse, the Guinea worm, and that thread-worm of the blood (*Filaria sanguinis hominis*) which afflicts man throughout the tropical zone. These are the INDIGENOUS enemies of high forms of animal life—the plagues of definite home, but occasional wide-spread fatality. Directly or indirectly, through the flora and fauna of a definite region, they spring from the soil, can always be traced to a natural habitat, and may properly be termed AUTOCHTHONOUS.

Lastly, we have to deal with maladies due to individual predispositions and to accident, isolated sickness, disease, or injury. A man suffers from dyspepsia, a woman from cancer, a child from rickets, or a horse from colic. In these and similar instances there is no pretense for confusion with contagious and with localized or indigenous diseases. The maladies or the individual instances of the maladies are INDEPENDENT and purely sporadic. These may be and are often due to special idiosyncrasy or to hereditary transmission.

Well do I remember the days when the doctrine of spontaneous generation was so facile an interpretation of all human ills, that the strict definition of a specific animal poison, and its peculiar history as an agent of destruction, met with great incredulity among physicians and others. Maladies such as smallpox or cholera were supposed to originate anywhere.

The tendency now is to the other extreme, viz.: that no widespread pestilence can occur without a definite disease-germ —a link between a preëxisting case and the present one. The well-informed in pathology run no risk in confounding relapsing or famine fever with such a disease as variola. When I first demonstrated the exotic character of the animal contagia in the British Isles, and pointed out to some of the best European authorities that epizoötic aphtha, the Russian murrain, the lung-plague in cattle, and genuine exanthemata, had no home, but traveled in the lines of communication established by wars or trade, and were never (in the popular acceptation of the term) acclimatized anywhere, the conclusion was deemed far-fetched and was only tardily accepted. For nearly thirty years my mind has been attracted to these questions, and, in all probability, the wider field of comparative pathology gave me no small advantage over those whose sole field of observation related to diseases of man. For the past ten years my attention has been directed to disinfection and disinfectants, and more especially to antiseptic cold. The last has led to that all-absorbing branch of practical engineering and applied physics, artificial refrigeration ; and little did I think, when I aimed at isolating the British Isles from cattle contagia, by favoring the importation of dead meat instead of live animals, that my researches might lead me, as I believe they must, to the practical demonstration of the means whereby yellow fever may be held permanently in check, if not totally extinguished, on the Atlantic seaboard.

These words of explanation are needed as an excuse for, perhaps rashly, forcing myself before the public with an exposition of views so widely different from those generally held. The public (and I include scientific men) is not satisfied with the current knowledge on the subject of yellow fever. There are many who scout the idea of informing the public on such questions. I belong to that class of ardent believers in popular knowledge who consider no adage more ridiculous than that which avers that a little "knowledge is a dangerous thing." Better a little than none, and that little will grow. The germ of truth remains, the lie vanishes.

We have arrived at a stage when we may generalize with profit, when at all events we may define, and I venture to believe that the classification of all maladies under the three great heads—

International or Contagious,
Indigenous or Autochthonous,
Independent or Individualized—
will clear away some of the mist enshrouding this great question ;
and in due time the skillful ætiologist—the student of disease-
causation—will split up these great groups into lesser and well-
defined orders, genera, and species, and abolish the ancient no-
sologies based on theories, swept away together, in due time, by
the torrent of medical progress.

Remember the three I's, and one of the fundamental features
of yellow fever can not be forgotten, viz., that it belongs *not* to
the contagious diseases. Its poison has not been collected on a
pin's point to propagate the malady from the sick to the healthy.
The virus of small-pox is known to us by its effects, and so is
the poison of yellow fever ; but there is not one of the pure contagia that will not yield a secretion—such as the system does not
produce in yellow fever—which can be desiccated or otherwise
preserved ; and so long as it is undecomposed, or retains its latent
vitality, when introduced into the blood of the susceptible crea-
ture, a disease is induced with a definite period of incubation, a characteristic invasion, and aggravation of symptoms,
until a *crisis* occurs, to be followed by death or convalescence.
So concentrated or so virulent may be the poison, that the
ordinary course and symptoms abort, and a rapid poisoning
results in what the French so graphically name *un accès fou-
droyant.*

Yellow Fever indigenous.

Yellow fever pertains to the indigenous plagues, the diseases
of localized origin ; and future observers may and no doubt
will, as past ones have done, ally with it, or show some rela-
tionship to it, of diseases in various parts of the world. Its
most striking and constant characteristic is development in a
confined atmosphere on or near the equator in the Atlantic,
with a virulence almost proportionate to the protracted heat,
humidity, and stillness of the air breathed by man. These are
the conditions under which decay or common putrefaction of
organic matter occurs ; and it were well to investigate further
the statements concerning the rapid decomposition of meats
and other perishable provisions in yellow-fever-laden air.

Damp Ships unhealthy.

Dr. Turner, Secretary of the National Board of Health, has written very forcibly on the insalubrity of damp ships—of ships whose air is kept damp by perpetual swabbing and washing. The conditions he describes and denounces are the conditions favorable to the development of what is called " ship fever," which in the tropical Atlantic means, to all intents and purposes, Yellow Jack.

The region whence yellow fever travels northward is remarkable for humidity, and that humidity does not depend on rains, which so often favorably influence the health of infected cities in the United States, but saturation of the atmosphere at high temperature—a state favoring the rapid rusting of metals and instant corruption of organic remains.

" Dr. E. H. Barton, of New Orleans, informs us, as the result of an examination of fifteen epidemics, that the average dew-point at the commencement of the sickness was 95·82; the average at the maximum of the epidemic was 74·34, and that at the declination 62·12. The difference between the period of the commencement and that of declination was 13·70; and we have the remarkable fact that a diminution of the dew-point brought the epidemic to its maximum of intensity, while a further diminution brought it to its declination. Dr. Barton remarks, in reference to this matter, that during the last nine epidemics it was found that, although the results stated were but an average of the whole, the extremes or variations from it, in any year, were very small; in other words, the minimum dew-point at which the epidemic passed off, and which was required to destroy that character, was noted at 58·26, and the maximum, in any year, under which it ceased its ravages, was 66·64, being a difference of a fraction over 8° only."*

Pending researches by a competent commission at Havana, it may be rash to predict that little more is likely to be learned of the specific germ of yellow fever than we know of the organic elements incident to and connected with putrefaction. There is a question of intensity, as well as of specificity, pending—an intensity bearing almost a definite relation to high

* La Roche, vol. ii., p. 159.

temperature and atmospheric humidity, darkness, and intestine inertia in the deepest holds of vessels.

Whatever may be the results of this year's investigations, I am sure that the concentration of attentive thought and skillful inquiry, now so fortunately inaugurated, will dispel much mystery, and especially any mystery still lurking, which surrounds the spontaneous or imported character of the disease, so far as the United States are concerned. The facts recorded during the past century already constitute a broad and inspiring basis on which to found a rational and perhaps incontrovertible opinion. The very discrepancies brought out by the contagionists and anti-contagionists—by those in favor of the theory of spontaneous development in the Southern States, and those holding the opposite view—have enlarged our sphere of vision.

We must not expect a prompt unanimity of opinion. That we may never have, but this off-hand essay would not have been written had I not, in the course of my inquiries with a view to the prevention of yellow fever, found so much to guide our thoughts into new channels, that I felt impelled to take others promptly into my confidence, with a view to expediting the experimental work which, in my opinion, must lead to the final and permanent exclusion, as an epidemic or pestilence, of ship or yellow fever from all lands.

Where do we find Yellow Fever? and whence does it reach us?

These two questions can undoubtedly be answered, and in the main answered conclusively. I shall not attempt to lay all the evidence in my possession before my readers. That would defeat my object, which is to be brief and prompt. I shall confine myself as much as possible to adducing those striking facts and data which seem to me established beyond question.   .

" Yellow fever was not known to the people of the Eastern hemisphere until after the discovery of America by Columbus. The earliest epidemics of which we have any historical information occurred during the first half of the seventeenth century, in the West India Islands." *

---

* " Conclusions of the Board of Experts authorized by Congress to investigate the Yellow Fever Epidemic of 1878." Washington, 1879.

It appears that yellow fever first occurred within the present limits of the United States in 1693, in Boston, Massachusetts, and in 1699 in Philadelphia, being imported both times from Barbadoes. It entered New York from St. Thomas in 1702, and for the first time reached New Orleans in 1796.

We must eliminate all doubtful cases. The recognition of isolated or sporadic cases (so called) of yellow fever is attended with great difficulty, and the most important point for future observers to clear up is the positive recognition in the earliest stages of characteristic signs of the disease. But so many epidemics have been studied that the mass of authenticated outbreaks is enormous, and some able physicians have arrived at conclusions which bear the stamp of ample justification. "I remain convinced," says Dr. Faget, of New Orleans, in his admirable letters on this subject, "that yellow fever never occurs in New Orleans unless it is imported; for in my opinion it owes its origin always to special germs (*sémences spéciales*) which are brought to us in the holds of certain ships." *

Yellow and Paludal Fever.

Dr. Faget published his "Mémoires" in great measure to define the contrast between ordinary marsh, miasmatic, malarial or paludal fever, and yellow fever; and he says: "Now the principal object which I am pursuing in my studies of the two fevers is that both can be attacked in their origin (*dans leur causes*) or their morbid principles; *the yellow fever* by special means of sanitation applied to the holds of importing vessels, there where the germs exist; *the paludal fever* by turning our marshes into healthy lands, and meanwhile, by quinine, pursuing the poison itself and destroying it, even in the depths of the organism."

Protection of New Orleans.

Dr. Faget remarks that after a winter with frost New Orleans can be readily preserved from all yellow fever, and at the time he was writing (June, 1863) five years had elapsed without a visitation of the disease. " The blockade, and, after the blockade, the severities of the state of war, that is to say, *the absence*

* "Mémoires et Lettres sur la Fièvre Jaune et la Fièvre Paludienne." Par le Dr. J. C. Faget. Nouvelle Orléans, 1864.

of maritime commerce [the italics are his], have protected us
marvelously, and especially in 1862, when our city was crowded
with thousands of strangers from the North, shut up (*entassés*)
in hot barracks, which were ventilated with difficulty. Moreover, in the winter of 1862–'63 we had a severe frost."

Original Views on Contagion.

Such facts, and all the facts known concerning the early invasions of yellow fever in Europe and America, at first led to
the unanimous opinion that it was a contagious disease. Devèze, Nathaniel Potter, Charles Maclean, and others successfully
combated this view. One of the benefactors of the human race,
Chervin, contested the prevailing opinion, and visited America
to collect facts in opposition to the doctrines propounded. In
1828, at Gibraltar, Chervin attributed yellow fever to the cess-
pools of the city. Louis and Trousseau inclined to the importation theory. There was so much truth in Chervin's views as to
the actually non-contagious character of yellow fever that, although he went too far and thought the disease might develop
anywhere, he was the great reformer of the rigorous and inhu-
man regulations of the French quarantine system, and received,
in recognition of his eminent services, the prize of 10,000 francs
from the French Institute.*

The two leading opinions which are fought over to this day
relate to communication by *actual contact*, contagion, or communication by *aërial infection*, once the disease manifests itself. At the present time I believe I am correct in declaring
that the statement made by the Board of Experts (*loc. cit.*) is
accepted by the best informed, viz.: "In the dissemination of
yellow fever, atmospheric air is the usual medium through
which the infection is received in the human system." It is,
however, not "carried to any considerable distance by atmospheric currents."

Yellow Fever defined as Nautical Typhus.

Faget, discussing the many theories respecting the nature of
yellow fever and marsh fevers, says, in a letter written in July,

* Faget, *loc. cit.*

1859 : " Audouard seems to me to have come nearest the truth in establishing a distinction between *nautical typhus* (yellow fever) and *paludal typhus* (malarial fever). Nothing seems to me to throw more light on the study we have undertaken than this remarkable distinction. If I do not err, it will be more profitable to enter on this course, which has been little studied, than to continue the discussion of *contagion* or *infection* which turns up whenever we have to deal with the question of *importation.*"

Chervin, the enemy of quarantines, said they would cease the moment it was proved that yellow fever was not contagious, and therefore could not be imported. Chervin accomplished a great work, and yet he was, as mankind usually has been, only half right. We blunder into wisdom. We fight and slaughter each other into enlightenment. He was perfectly correct that yellow fever is not a personally contagious malady like small-pox ; but he overlooked the fact that the common putridity of a foul city may harbor and nourish the uncommon putridity resulting from imported infection in ships.

Obstructive Quarantines.

We always fly to extremes in moments of panic. Good government demands the prevention of frantic popular terror, by the investigation and publication of truth in relation to all epidemics, as much as by the firm adoption of preventive measures against disease. Dr. Vanderpoel, of New York, recognizing fully the imported character of yellow fever, raises his voice, as Chervin did, against oppressive and *obstructive* quarantine. He says : " The pestilence of the past season, which has raged so fearfully through the Southwest, has produced a feeling akin to terrorism throughout that whole section. The leading physicians make no other suggestion against future invasions of yellow fever than absolute *non-intercourse* between that section of country and the West Indies for six months of each year." *

At the present moment (August, 1879), throughout the South, the people are in a state of abject terror, flying from their homes, and tortured by a stalking specter—the jaundiced

---

* " Quarantine, with reference solely to Seaport Towns." By D. Oakley Vanderpoel, M. D. New York, 1879.

image of death. Shot-gun quarantine and the brutal repulse of wanderers are the evidences of an inhumanity engendered by fear. The Memphis outbreak should be the last of these terrible inland visitations. Even Chervin recognized that the foci of infection were derived from vessels at sea, true ambulating foci (*vrais foyers ambulants*), which constitute the radiating points whence the disease extends more or less widely among the adjoining dwellings. He recognized that the ships were charged with the infected atmosphere, the *potential cause* of yellow fever outbreaks.

Speaking both of yellow fever and cholera, the Board of Experts said: "It is now known that the poison of neither is susceptible of long vitality when exposed to the open air; but it is not yet determined how long its infectious properties may be preserved in closely shut chambers or in compartments of vessels, or when inclosed in the folds of clothing or goods. It consequently follows that ships are especially dangerous carriers of these diseases, and also that they remain sources of infection for months after having been infected with the poison."

Yellow Fever in Steamers and Sailing Vessels.

Dr. Vanderpoel has contributed, in his work on quarantine, some valuable observations on the relations which steamers and sailing vessels maintain to each other, in disseminating the germs of epidemics. He says:

"First, steamers are far less liable than sailing vessels to become infected with the germs of yellow fever. It is admitted by most writers that yellow fever is not contagious, or, if under certain conditions it becomes so, it is so slight as not to require consideration in this connection. It is then transmitted from place to place by the vessel, the cargo, the baggage and effects of passengers and sailors. The conditions which would modify this receptivity in the two classes of vessels should therefore be taken into account. Of the ways of transmission, the vessel itself plays the most important part. The germs of yellow fever increase with fearful rapidity where the conditions of filth, heat, and fermentation are in active operation. These are supplied in full measure in the hold of a vessel lying for weeks under a tropical sun, being loaded, as most of those coming from fever ports are, with crude sugar and melado. It is in the bilge of the vessel where this filth accumulates, and where the fever-germ revels in its propagation. The facilities for removing this filth and maintaining cleanliness are far greater on steamers than on sailing vessels. In the former, in addition to the filth from the cargo, the oil and dirt from the machinery and the water from condensed steam settle in the bilge, to remove which the steam-pump is brought into frequent use. Not only can the matters which accumulate there be readily pumped out, but clean water may from time to time be thrown in to remove more effectually the filth, and preserve the vessel from disagreeable odors. On the other hand, in sailing vessels all pumping must be done by hand, and is just so much less effectual and imperfectly performed; consequently, in the same proportion is the vessel liable to convey the morbid germs.

"Again, steamers, in most instances, belong to lines which run at regular intervals, remaining in port but a few days, as the cargo is already engaged by regular consignees. Sailing vessels, on the other hand, often lie for weeks in an infected port, under a tropical sun, and their officers are too often indifferent to the hygienic conditions of the vessel and crew, until the vessel becomes a pest-house of sickness ; or, if the occupants escape while still in port, the germs remain in the bilge, and yellow fever breaks out later—the length of the voyage, a fermentable cargo, a dark and heated hold, all favoring the development of the disease.

"As just intimated, the length of time occupied by the voyage is a decided element in favor of steamers, in lessening the opportunity for the development of the germ ; for, as I shall show when speaking of the measures proper to repress yellow fever, I regard it of the highest moment that vessels should be discharged as soon as they enter port. Steamers also possess the appliances for forced ventilation, and currents of air may thus be kept circulating throughout the vessel, a condition unfavorable for germ-propagation. Experience of some years has, I think, shown that steamers rarely become infected ; while not unfrequently they reach this port with sick persons, it will usually be found that these individuals came on board during the incubative period of the disease, having contracted it on land.

" The average duration of the trip of a sailing vessel extends in most cases far beyond the period of incubation, and the cases of sickness she brings to port must be regarded as *deriving their origin from the vessel.*

" While as a precautionary measure both must discharge in quarantine, the deduction is a fair one that far less danger of importing the disease is to be ascribed to the steamer than to the sailing vessel, and that measures of cleanliness and disinfection should be more thoroughly carried out on the latter. It may also be added that the facilities of cleaning with steam-pumps, after discharge of the cargo, are manifestly also in favor of the steamer. It is also a proper question to consider whether an iron vessel would be as likely to become infected as one of wood, where the saturation of the latter would be another element in the fermentative or decomposing process.

" While, then, I regard steamers as exerting a favorable sanitary influence in the prevention of the transmission of yellow fever, as compared with sailing vessels, their relation to the transmission of cholera is the reverse. This latter disease is transmitted primarily by the individual, and not, as in yellow fever, wholly by the surroundings."

What Ships may be considered infected.

A ship entering a seaport infected, or developing in the waters where it is lying at anchor a yellow-fever infection, has almost always been charged with contracting the disease from land. Even a superficial study of certain West Indian islands indicates that the chances of contamination by organic decay are infinitely less on shore than in the shipping; and it has not unfrequently happened that a foul vessel has discharged apparently healthy passengers and crew, and, on hiring labor to clean the hold and purify the bilge, the susceptible landsmen have been instantly stricken with fever.

Every ship therefore, whose sanitary condition is doubtful, hailing from the West Indian seas, without the stain of having communicated with an infected port, may be regarded as possibly or probably infected.

Epidemic Nucleus adjoining Shipping.

The extension of yellow fever from the shipping occurs usually to dwellings adjoining wharves; or the sailors and passengers from an infected ship crowd with their baggage and clothing, which are tainted by the air of the hold, into close quarters, where an impure atmosphere speedily multiplies the putrefying centers of effete organic matter. So near may be a foul ship to a readily fouled dwelling, that, with a favorable wind, the pestilential vapors find their way on land with fatal effect, without any human carrier; but the distance that infection can be usually carried, by the unaided wind, is exceedingly small.

Dr. Faget has referred specially to two epidemics which he studied closely in New Orleans, viz., that of 1853 and that of 1858. He shows how the "epidemic nucleus"—the center of origin—defined itself promptly in 1853 around a circumscribed part of the port at Lafayette, precisely where certain cases of yellow fever had been recognized in the shipping. In 1858 the "epidemic nucleus" was at the end of the street near the pier 33, where public rumor had indicated the presence of yellow fever among the ships.

Course of the Yellow Fever in Cities.

From the original centers of infection, the most accurate records indicate the spread of the disease, whenever it has been newly imported, from house to house and street to street, proximately, without intervening healthy oases, steadily invading in advancing outline the infected area, as oil spreads when dropped, as Dr. Faget graphically puts it. One part of a city is the seat of the virulent epidemic, and another is free. The doctors in one part are prostrated by overwork, whereas in another idleness and indifference prevail, until the infected atmosphere has crept from house to house and street to street, drowning in fetid vapor every habitable quarter. At last it meets a barrier; it reaches the city limits. In the country, beyond the afflicted town, cases of malarial fever and local black vomit occur at all times, but yellow fever never, except among refugees. It is here that the competent observer can appreciate the immense difficulties of diagnosing isolated cases of yellow fever, and the ease with which the course of the malady may be traced from infected centers to the edge of the surrounding country, into which it can not penetrate as an epidemic.

Dr. Faget has brought this peculiarity of yellow fever invasions out strongly and clearly. The learned Dr. Francis says, referring to its importation and extension in foul cities, "impurities can not change its nature, but may augment its influence."

Location of Infected Cities.

Dr. Daniel Osgood, of Havana, in a letter on yellow fever in 1820, says: "Adjacent to all the places in which this fever becomes endemic is the sea, or some other body of water. In the West Indian islands, with the exception of the very small ones, persons who reside in the inland parts, when affected with fever, have it not of this kind. But instances of its being of this kind are seen in them, when they have become disordered shortly after they have left the interior and removed to the sea or a harbor. On the contrary, even strangers from cold climates, who on the first arrival at a tropical one go into the country and remain there, escape the disease."

In confirmation of Dr. Osgood's opinion, the Camp Jacob in the interior of Guadaloupe has been established for fresh troops from France, so as to acclimatize them. On the coast of Africa we observe exactly the same conditions. It has occurred at St. Louis, Cape de Verde, Gorce and Bissagos Islands, Sierra Leone, Cape Coast Castle, etc., but never beyond the points of direct communication with shipping. No land in Africa breeds it.

And here is the strong point which, in Dr. Faget's writings, confirms the opinion I had formed when Dr. Woodworth first consulted me about ship disinfection. He says distinctly that no cities attract yellow fever, unless on the seashore, or in close and direct communication with the commercial marine.

Every city in which ice forms freely in winter is thereby purified from yellow fever; and, according to Dr. Faget and others, no disease can occur there subsequently except as the result of reimportation. In those cities, like Havana, Vera

Cruz, and the Brazilian ports, where frost is unknown, susceptible persons may contract the malady at any time, especially in
summer; but these ports might again become pure were the
shipping not permanently infected and eminently fitted to keep
up the reproduction of a sufficiently virulent poison to perpetu-
ate the malady.

Is it not singular that, in the immediate vicinity of Vera
Cruz, such a port as Tampico should be more rarely infected,
and only when contaminated by the shipping; and that a town
almost adjoining Vera Cruz, like Jalapa, should never be visited by the malady except among infected refugees from Vera
Cruz? So long as the heat and stagnant air permeate every
house in Vera Cruz, it is most dangerous, even for transient
travelers; but toward the fall of the year a "norther" venti-
lates the dwellings, and infection is rare.

Dr. A. N. Bell, in an ably written paper in the "Transactions of the Epidemiological Society" for 1867, says: "Stagnant
air, dampness, darkness, and warmth are frequently the insepa-
rable conditions of vessels in warm climates. If to these conditions there be added a filthy vessel, putrid provisions, or bad
water, or if the cargo consist of materials peculiarly liable to
infection, such as rags, hides, feathers, sponge, or sugar, the circumstances are then complete, not only for the reception of the
poison, but for its origination and continuance," in the delta of
the Mississippi, which Dr. Bell considers the home of yellow
fever.

Another and an older ship-surgeon, in a report published in
the "Medico-Chirurgical Review" for 1840, says: "That in
many cases, and the worst cases, West Indian fever is essentially
connected with some agency in the interior of ships, altogether
independent of personal communication, or collections of extra-
neous matters, all discriminative experience shows."

Briefly, yellow fever is a disease of the intra-tropical and
juxta-tropical Atlantic ocean, always infesting ships, and only
attacking towns in direct communication with those ships; dis-
appearing entirely whenever frost can reach it; but continu-
ing almost perennially in some cities which are in constant intercourse with shipping, and in which the temperature is never
low enough to destroy the poison.

Altitude as affecting Yellow Fever.

Yellow fever is a disease of a dead sea-level; and, the deeper the ship and flatter the city, the more readily it pullulates and destroys. Gravitation favors stagnant vapor and molecular masses of organic decay. Below the sea-level in the holds of ships its maximum intensity is matter of history. In low, close dwellings and cellars on land, the mortality is frightful. Altitude bears an indirect ratio to the prevalence of the malady, all other conditions being equal.

Since, as we have stated, the malady does not spread beyond inhabited seaports in low situations, and is intimately associated with confinement of human beings in inclosed spaces, if cases of yellow fever have occurred, as Humboldt says, 3,243 feet above the sea, it could only have been exceptionally among persons who have contracted the disease near the sea-shore or accessible shipping. Memphis is 400 feet above the sea-level, and there is no doubt as to the virulence of a genuine epidemic at this altitude on the banks of the Mississippi; but, the higher the land and the drier the atmosphere, the more difficult is the propagation of the disorder.

Barometric Pressure.

Dr. Barton, of Louisiana, demonstrated, so far as his observations extended, a relation between the prevalence of yellow fever and high barometric pressure. Basing his calculations on an examination of fifteen of the epidemics which had occurred in that city, the mean pressure of the atmosphere was, at the commencement, 30·108; at the maximum of the epidemic, 30·024; and at the declination, 30·074. The difference between the pressure at the commencement and at the maximum of the epidemic is ·084; and that between the maximum and the declination, ·050.

Is there a Proper Soil for the Disease?

The learned Noah Webster,* not a medical man but a pure historian, set himself to prove at the close of the last century that celestial, telluric, and volcanic disruptions and other pro-

* "A Brief History of Epidemic and Pestilential Diseases," Hartford, 1799.

digious phenomena cause epidemics; that "the weapons of Apollo," translated into "intense heat of the sun," and Diemerbroeck's *seminarium*, or deadly pestilent germ, sent from heaven, developed all plagues; that the Hippocratic *to theion* and *seminarium e cœlo demissum* was the yellow-fever germ of the present day. Boldly does Webster declare: "That some such general cause exists in the atmosphere, at certain periods, will be rendered very probable, if not certain, by the facts hereafter to be related."

As easily as he noted coincidences between comets and boils have others recognized an indispensable affinity between yellow fever and land; first all lands, and then wisdom has led them to discriminate against other lands but their own. Dr. Francis vehemently uttered: "No land has been more earnestly libeled by her own sons than ours by our own who have adopted the domestic theory."

A navy surgeon, who had ventured to draw a sound conclusion from that which he had seen, says: "The yellow-fever poison may originate on shipboard independently of any communication with an infectious city." It is to be regretted that Dr. Wilson, who penned this phrase, should have indulged in the belief that Nature is uncertain and irregular in her ways; that so simple and sufficient a cause should have to be supplemented or modified by something else. He says: * "The yellow fever, prevailing more or less constantly in all the cities of tropical America, is endemic in those cities. When unusually prevalent and exceedingly fatal, without apparent reason for this change of character, yellow fever is said to prevail epidemically, or to have become epidemic. In this case an *epidemic influence* is supposed to be superadded to the ordinary causes of the disease."

A sensible naval surgeon, whose words I have already quoted in another connection, but who is only known to us as a nameless reporter to the "Medico-Chirurgical Review," on the health of the navy from 1830 to 1836, says: "If this disease were the spontaneous production of America, how comes it that it did not destroy the British armies which acted in the late war in Pennsylvania, Virginia, and Carolina, as it has done of late in the West Indies ? "

* "Naval Hygiene," Washington," 1870.

And if we seek for any expression, on the part of well-informed authorities, in the West Indies, we utterly fail to learn anything of local conditions adequate to more than the development of periodic affections. In many cases we have evidence of the great salubrity of individual West Indian islands. Referring to St. Eustatia and Curaçoa, Dr. Peixotto published a vigorous defense in 1822, and described both as being dry, rocky, and proverbially healthy. He says:

"Curaçoa is situated in latitude 12° 8' N., and longitude 69° W., distant twenty leagues from the coast of South America, of which the highlands are plainly discernible in clear and calm weather. The island extends from the northwest to the southeast, in length forty miles, and its mean breadth may be fifteen. The general appearance of the island is barren and parched, its surface broken into abrupt masses of hill, which raise their picturesque forms in every direction. They are, however, bare of all marks of vegetation or verdure, cultivation being confined to the vales, in which the fruit and maize plantations are more advantageously located.

"The soil is superficial, consisting of a slight mold covering a rocky substratum, in which gypsum, pumice, ochre, and some argillaceous earth are found. Water is to be had by digging into the rock; but there are no rivulets of fresh water, nor any marshes to be seen in any part of the island. The hills are of no inconsiderable height, some of them having an elevation of from ten to fifteen hundred feet above the level of the sea. They no doubt serve a highly useful purpose, by collecting and conducting the currents of air along valleys; and, adding to their force and impetus, they refresh the atmosphere."

Abbé Raynal and Bryan Edwards, in their admirable works on the West Indian islands, describe no feature—no general or localized conditions—to which we can ascribe any but the ordinary malarial diseases. Indeed, Mr. Edwards, in describing the uniform high temperature in all the lands beneath the tropic of Cancer, the distinction between the two seasons, the wet and the dry, how the weather becomes dry, settled, and sultry when the tropical summer reigns in full glory, adds a note—a contrast between the healthiness of the West Indian islands as compared with the southern provinces of North America. "In Virginia,"

Mr. Jefferson relates, "the mercury in Fahrenheit's thermometer has been known to descend from 92° to 47° in thirteen hours." And, with pride, exultingly he adds: "The West Indian islands are happily exempt from these noxious emanations," produced by sudden change from great heat to great coolness.

The most diligent search will fail to reveal that even in their ancient state they were noxious to health. Maize was regularly sown twice a year in the plains or savannas; the islands were populous and necessarily cleared of underwood; the trees which remained afforded a shade that was cool, airy, and delicious. "Such were these orchards of the sun and woods of perennial verdure," says Edwards.

The sanitary history of the West Indian islands, after Cornilliac, is sketched in the next chapter, and we derive from a general survey the strongest presumption that not even Martinique and Cuba, which have suffered most, present a single rood of soil in which yellow fever was ever known to develop spontaneously. If it has, the crop has been most erratic, and unlike the unhappy perennial results of man's abode in swamps and jungles.

Some of the cities now constantly preyed upon by yellow fever, such as Rio de Janeiro, were absolutely free from the malady not many years since. Rio was, for over a century probably, as free as any European port; but commercial intercourse has given it the *tropical ship fever*, the permanent source of contamination being ships on which yellow fever infection often continues so long as such vessels float.

The industrious La Roche, compiling with the greatest care, writes with a mild bias, sufficient to show that his mind was not free from views which time must modify. He says: "Be the causes, however, what they may, on one point there can be no doubt—that the yellow fever has geographical limits, beyond which it does not appear, and that within those very limits there are many places where its usual apparent cause would seem to exist, but where, nevertheless, it has never shown itself, or has done so very seldom. *The West Indian islands and part of the coast of South and North America constitute its proper soil.* From Brazil to Charleston in one direction, and from Barbadoes to Tampico in another, the causes of this form of fever are in constant, though unequal force, in regard to different seasons and localities." *

"Its true area," says La Roche, "includes the Caribbean and other islands called the West Indies, and Bahamas; the contiguous coast of Colombia and Guatemala," northward as far as Boston. The true area is partly around the regions indicated in this quotation, and not near the Atlantic shore of the United States, embodied in his full sentence, and which I embrace in the words· "northward as far as Boston."

The Board of Experts say : "From the West Indian islands, as the earliest historical focus of the disease, yellow fever has been carried at different times into several of the countries of Europe, Africa, South America, and North America.'

" Yellow fever has never made its appearance on the great continent of Asia, nor in Australia, nor in any of the islands of the Pacific Ocean ; and it has not prevailed extensively on the Pacific coast of the American continent.

" In all the countries outside of the West Indies which have been visited by it, yellow fever is an exotic disease ; and in all of them its introduction can be traced either directly or indirectly to the West Indies. In some of them it seems to have established itself permanently, and to have become endemic ; as, for example, in the Brazils. In most of them it has failed of naturalization, and successive epidemics can be traced to successive importations."

Dr. Daniel Osgood, of Havana, shows how the West Indian islands can not be the "proper soil" for the disease now, if they were not the " proper soil " before the days of Columbus. It was not seen, wherever it is said to have arisen, until after a considerable progress had been made in the population of those places, covering town sites with streets and buildings, and establishing centers of active trade.

Yellow Fever an Exotic in every Land.

The malady is as much an exotic in the West Indies as it is in New Orleans. In relation to all lands, I say what Dr. Francis said of New York: "The disease is an exotic . . . . in

---

* " Yellow Fever," by R. La Roche, M.D., Philadelphia, 1855, vol. i. (p. 119).

all cases an imported pestilence." Havana has no greater affinity for yellow fever, or the fever for it, than Pensacola; and it behooves the British, Spanish, Mexican, and Brazilian Governments, as well as the United States, to join hand in hand firmly and speedily to test the truth of such an assertion. Valuable as the scientific researches at present in progress must prove, we have here work demanding less of microscopic investigation than broader ætiological, meteorological, and geographical study.

The impression produced on my mind from the first by the conclusions of the Board of Experts was eminently favorable. Reading them, after months of thought and greater familiarity with the question, I still consider them in the main sound; but on some fundamental points, when yellow fever is practically extinguished by the best known means of sanitation, those conclusions, prepared in 1878, will mark a line of distinction between the matured views of the period and the revelations of subsequent inquiry. The most obvious conclusion arrived at by the Board of Experts is that yellow fever is a disease of cities; that its habitat is the West Indian islands; and that it has not failed of "naturalization" in "countries outside of the West Indies." From this I unequivocally dissent.

Naturalization of the Disease on Land impossible.

Ague is a disease natural to marshes, and the splenic fever of Texas is peculiar to a certain soil and grass. We have no difficulty in migrating to or from the regions where these diseases exist, but no such knowledge is possessed in relation to yellow fever. Popular opinion has long adopted the view that any malady afflicting a people may in the long run so establish itself among them that it can not be eradicated. Such a view is totally erroneous in relation to the pure contagia and to plague or yellow fever: they cease when the active causes of their development cease. The dependence of some diseases on the aquatic organisms of a country is well known to medical men. The alternate generation of trematode or sucking worms demands water and animals that live in water. The earlier and more immature forms of animal life occur in water. In the air they dry and die. The rich, warm, and phosphorescent waters

3

of the tropical Atlantic have great capabilities in favoring certain forms of putrescence. Many animal and vegetable forms in them die so soon as they are inclosed where stagnation and decay may be favored. This putrefiable matter abstracts oxygen from the sulphates, producing sulphurets; a sulphuret of sodium is formed; this again, acting upon the water, decomposes it, and sulphuretted hydrogen is disengaged. No sulphuretted hydrogen is found in sea-water, even on marshes, where an abundant oxidation is possible. Fermentation or putrefaction is essential to this complex chemical decomposition. A slight access of air is necessary to start it—free ventilation kills it. The deoxidizing power of putrefying substances is very remarkable, and the irrespirable character of gases rising from the bilge can be readily accounted for.

Its Propagation.

The remark made by the Board of Experts, that yellow fever has never appeared in any place in the United States in a shorter time than would enable intercourse with an infected place, is correct as applied to inland dispersion of the disease, but it is quite certain that the disease has frequently been generated or developed in a ship at sea between two ports, or in some cases possibly within the harbor of the city it has infected. This depends on latitude ; and, while no such origin *de novo* could occur in the waters of New York, it certainly could in Florida straits or even Havana harbor.

"Yellow fever is transmitted through the interior of a country by steamboats, barges, and other river craft ; and by railroad cars, wagons, carriages, and other land vehicles ; and the infection may be attached to the boats, cars, or other vehicles themselves, or to their cargoes, or to the persons traveling upon them with their baggage." *

Alternating between Sea and Land.

Mr. Tytler, in 1799, said : "This distemper attacks sailors in the West Indies more than any other set of men, even of new comers." Dr. Stevens saw cases of yellow fever in St. Do-

* "Conclusions of the Board of Experts," pp. 15, 16.

mingo during the months of August and September, 1793; and these were entirely confined to American seamen, while the native inhabitants of the city were totally exempt from it. It is well to remember that the malignancy of yellow fever, as Dr. Bell has shown, appears to depend primarily and most generally upon suddenness of exposure to the poison, and secondly upon an atmosphere unfavorable to recovery. The islanders, accustomed to the climate, have acquired a certain degree of resistance to all heat disease. The sailors arriving in the dangerous tropical season from healthy latitudes are exceedingly susceptible; and, from the confined space in which they are compelled to live and sleep, they are exposed to a concentrated infection.

The conditions on shipboard and in a house are not dissimilar, *except in degree*, as to stagnation of air and accumulation of putrescible organic matter in the air and recesses; and there is abundant evidence of the element of time being in some sense related, in the ship and the house, in the development of foul air and yellow fever. Aërate and disinfect the house or the ship, and the disease is gone. It is in the true sense of the words endemic or naturalized nowhere, in no fixed spot; and on a clean coast it contaminates the seaport towns without reaching the villages of the seashore close by. There is an alternate transit from ship to house and from house to ship in seaports; but anywhere both can be, in my opinion, radically and permanently purified. I do not hesitate to condemn the too commonly used expressions *acclimatized*, *naturalized*, and *localized as an endemic*, in relation to any city on the Atlantic seaboard.

Humboldt's Views.

What did Humboldt say many years ago? "In all climates men imagine they find some consolation in the idea that a disease reputed pestilential is of foreign origin. As malignant fevers are easily engendered, amid a large crew, crowded together (*entassés*) in filthy vessels, the commencement of an epidemic dates pretty often from the arrival of a squadron. Then, instead of attributing the evil to the vitiated air contained in vessels deprived of ventilation, or to the effect of a hot and unhealthy climate on newly arrived sailors, people affirm that it has been imported from a neighboring port, where the vessels

in the convoy have touched, on the passage from Europe to
America. It is thus that we often hear it said in Mexico that
a vessel of war, in which such or such a viceroy has arrived at
Vera Cruz, has introduced the yellow fever, which had ceased
to reign for some years past; it is thus that, during the hot
season, Havana, Vera Cruz, and the ports of the United States
mutually accuse each other of being the source of the disease
by which they are visited."

Probable Origin in the Tropical Atlantic Basin.

I have shown that yellow fever undoubtedly belongs to the
second great group of indigenous diseases, but the conditions
favoring it are on the ocean; the major factor in its genesis
is traced, by the general history of yellow fever, to the region
of intense heat in the Atlantic, possibly to the whole equatorial
belt, from 35° north latitude across the tropic of Cancer, and
down to latitude 5° south. Its most prolific region is west of
25° west longitude, and between the calm belts of Cancer and
the Equator in the Western Atlantic. "The Amazon, always
at a high temperature, because it runs from west to east, is
pouring an immense volume of warm water into this part of
the ocean. As this water and the heat of the sun raise the
temperature of the ocean along the equatorial sea-front of this
coast, there is no escape for the liquid element, as it grows
warmer and lighter, except to the north." The waters must
flow north. The great equatorial caldron, "which Cape St.
Roque blocks up on the south, disperses its overheated waters
up toward the 40th degree of north latitude, not through the
Caribbean Sea and Gulf Stream, but over the broad surface of
the left basin of the Atlantic Ocean." (Maury.)

"Until recently," La Roche tells us, "the river Amazon,
which divides Brazil from Guiana, formed the boundary of the
disease south of the equatorial line ; for, although it is said to
have prevailed at Olinda from 1687 to 1694, and to have shown
itself as far as Montevideo in the beginning of the present century, the latter circumstance is open to some doubt ; while in
Brazil, from the close of the seventeenth century to the middle
of the present, the disease was not observed. Since 1850 it has

invaded Rio Janeiro, Bahia, Pernambuco, and other places of that country." *

The history hereafter given of Anson's voyage round the world indicates that in 1740 St. Catherine's Island, in 27° south latitude, was visited by yellow fever of the most malignant kind; but, so far as we know, the disease was limited to the British squadron. Anson's ships sailed west from Madeira, became sickly in the equatorial belt, and struck south on nearing the American continent.

Usual Yellow-Fever Course from the Tropics northward.

These points in relation to temperature and equatorial currents may account for the localization of yellow fever in this region, and almost justify the old name of typhus tropicus. The past history of yellow fever indicates that the disease travels in the direction of the northward flow of the heated Atlantic and the Gulf Stream rather than southward.

" A distinguished physician of New Orleans, Dr. E. H. Barton, to whom the scientific world is deeply indebted for ample and correct meteorological records and valuable observations relative to the etiology of the yellow fever, has called attention to the fact that within the limits of the zone above pointed out the disease usually commences its epidemic career in the south, and progresses gradually toward the north." He points out the course and cities successively infected from May to September in any year, and adds " that in this mere historical statement it is not intended to be implied that the yellow fever is imported from the south to the north in this regular gradation, but merely that the physical changes inviting and producing its development become evolved as the season advances."

Are not, however, the following words of La Roche highly suggestive, when we connect the origin of the disease with the ocean rather than with cities where the disease is supposed to have become *naturalized ?* He says it follows " that it usually commences earlier in southern latitudes, inasmuch as the thermometer reaches there the requisite average sooner than in the northern sections of the zone. But, as we proceed, facts will be

* La Roche, op. cit., p. 120.

adduced which show that the disease has not unfrequently appeared later in southern latitudes than it usually appears with us."

Floating Center of Origin.

Its center of origin is a floating one, and the combined factors, incident to human intercourse in the ships and cities of the tropics, skip and scatter the undulating plague, in its yearly incursions on land, to the confusion of past history, the dismay of all, and the destruction of terrestrial endemicity dogmas.

La Roche tells us that, "according to Vines, it appeared in Barbadoes in 1647. Ligon, who records the same fact, states that the inhabitants of the island, and shipping too, were so grievously visited with the plague (or as killing a disease) that, before a month had expired after his arrival, the living were hardly able to bury the dead. In speaking of the following year, 1648, Du Tertre says: ' During this year, the plague, unknown in these islands since the time they were inhabited by the French, was introduced therein by some vessels. It commenced at St. Christopher, and in the course of eighteen months carried off one third of the inhabitants.' Writing ten years later, Rochefort remarks : ' The plague was formerly unknown there (the West Indies), as well as in China and some other eastern countries. But a few years ago, the greater number of these islands were afflicted with malignant fevers which the physicians regarded as contagious.' It prevailed in Jamaica in 1671. The event was connected with the return of the victorious fleet ' from the signal Panama expedition,' when ' they brought with them a high, if not pestilential, fever, of which many died throughout the country.' The fever appeared twenty years later at Léogane (St. Domingo), in 1691, on which occasion, according to Moreau de St. Méry, it was brought by the fleet of Admiral Ducasse. The same year it showed itself at St. Christopher, and the next at Port de Paix. The outbreak of the disorder, and its extensive ravages in Pernambuco (Brazil), from 1687 to 1694, is known to most readers ; so, likewise, its supposed introduction thence into Martinique in 1690, by the Oriflamme, which there touched on her return from Siam. It prevailed also in the city of the Cape (St. Domingo), in 1696, and at Léogane in 1698. Hughes, in his ' Natural History of

Barbadoes,' writing on the authority of Dr. Gamble, mentions the prevalence of the fever in that island in 1691. It was then and there called the new distemper, or Kendal's fever, after a distinguished and popular officer of that name. Hughes also mentions its occurrence in 1696. According to Captain Philips, who visited the island in 1694, it prevailed there very extensively, and had done so during the war some years before."

Long after yellow fever had first declared itself in the American Archipelago, its recurrence in seaports, sometimes at long intervals, led medical men to fix new dates and new sources of its origin. Thus Warren, who lived at Barbadoes in 1739, believed the disease had never been in the island before 1721, when it was brought from Martinique in the Lynn man-of-war. The disease appeared, according to him, a second time in 1733, when it also reached Martinique. He believed it to be of Asiatic origin, and ascribed it to a Provençal fleet which arrived from Marseilles at Port St. Pierre in Martinique, having on board several bales of Levant goods taken in at Marseilles from a ship just arrived from St. Jean d'Acre. Upon opening these bales of goods at Port St. Pierre, the distemper immediately showed itself, and well it might if they came from a vessel infected on the passage in the tropical Atlantic. The coincidence might be accounted for by the now known frequent attack of men engaged in unloading the vessels, and custom-house officers, who first have access to the hold of the ship after it has become infected on the high seas.

Yellow Fever originating in Mid-Ocean.

Sir Gilbert Blane, in his work on the "Diseases of Seamen," relates: "With regard to the effect of putrid exhalations, I need only mention that at the time of the battle of the 12th of April, 1782, there was not a sickly ship in our fleet; but many of the officers and men who were sent to take care of the French prizes were seized with the yellow fever; and it was observed that when at any time the holds of these ships, which were full of putrid matter, were stirred, there was an evident increase of those fevers soon after."

Dr. Copland, entertaining views not very dissimilar from those suggested before by Humboldt and Audouard, says that

since his visit to several places in Africa, and knowing the very
limited space in which a large number of slaves are often confined, both on shore and in slave vessels, he entertained the idea
that this pestilence or its *seminium*, or specific infection, had
been generated originally by the congregation of negroes in a
close atmosphere, or is generated *de novo* by this race when
placed in the circumstances now stated ; and that, although it
affects them in a comparatively slight manner, it is most particularly baneful to the natives of the colder regions. Of course
the slavers arrived near the equator reeking with foulness, and
readily imbibed the local factor of yellow-fever infection, as they
approached the eastern shore of Central and Southern America.

La Roche refers to records of the rise and spread of yellow
fever on board of ships occurring, *as they often have done*, under
circumstances which forbid the supposition of a foreign source
of contamination, and point to the vessel itself as the focus of
infection. He says: "A careful perusal of the numerous facts
we possess on the subject before us, so far from justifying the
propriety of discarding as spurious and erroneous all that has
been said in support of the malarial origin of the fever on shipboard, will fully sustain the opinion of those who ascribe, in
very many instances, the appearance and spread of the disease
in such localities to the operation of morbid effluvia generated
in the timbers of the vessels themselves, or the materials they
may contain."

Dr. Currie, of Philadelphia, complimented by La Roche as
"a distinguished countryman of ours," never could discover the
most remote reason for admitting that the yellow fever had ever
originated in this country ; he thought that "crowded transports
or ships of war generally, if not always, constituted the original
and proper sources of the matter of contagion or the poison of
the disease."

Dr. Faget states most categorically that facts prove that
yellow fever may break out in ships at sea (*même en pleine mer*),
without searching beyond them for any cause capable of producing such a malady.

Through the kindness of Dr. Cabell, the President of the
National Board of Health, I have had the gratification of reading a most able letter which appeared in the "Pensacola Ad-

vance" on the 30th of April last. The author of that letter
was Dr. Robert B. S. Hargis, and in it he says, backed by a
quarter of a century's experience : "After mature consideration,
I long since reached the conclusion that yellow fever is not only
exotic, so far as North America is concerned, but never originated *anywhere on land ;* and that it had its genesis on ships
or the sea-going vessels within the tropics." Further on he
remarks that "filth, either *per se* or associated with an epidemic
constitution of the air, free from yellow-fever germs found
abroad, will not cause yellow fever on land. I say 'on land'
anywhere on the face of this globular mass we live on." He is
explicit as to its undoubted origin. He says : "In the hold of
a ship, in which proper hygienic or sanitary measures have been
neglected, that has sailed within the tropics a considerable length
of time, there may, nay do, often exist powerful causes of infection of a peculiar character or nature, due to the sea water,
which always in the ship's hold contains large quantities of dead
organic matter, peculiar to itself, mixed with dead rats, and in-
sects which always infect ships to a very great extent, and which,
together with the *débris* of an assorted cargo, excluded from
fresh air and light, and subjected to a very high degree of heat,
exceeding often 110° Fahr., give rise to that very peculiar and
specific infectious principle that causes yellow fever. In the
hold of a ship, then, in the tropical seas, may be found the habitat of the yellow-fever germ ; there the *fons et origo mali* exists, which has been the 'stone of Sisyphus' to New Orleans for
three quarters of a century."

Dr. Hargis cites a number of cases on which his very positive opinion is based. He refers to the ship Le Bœuf in 1635 ;
in 1726, in the fleet of Admiral Hesser off Portobello ; in 1746,
the Spanish ships Angel and Astra on their way to the West
Indies; in 1783, in Admiral Solon's squadron ; in 1785, Spanish
ship Aldefensa ; in 1703, British ships Beaford and Kent ; ship
Hankey in 1792 ; besides ships in our own day.

La Roche accepts without qualification the opinions of the
origin of yellow fever on shipboard from local sources of infection, and independently of external influences. * He discards a

---

* "In the performance of this task, attention might perhaps be called to the
history of the disease as it prevailed on board the Princess Caroline, at Curaçoa, in

host of doubtful instances.   He refers to it the epidemic which
occurred on board the U. S. ship Delaware in 1790–1800 at
Curaçoa.   Dr. Anderson, of that ship, informed Dr. Vaughan, of
Wilmington (Del.), "that the disease unquestionably originated
on board, in the harbor of Curaçoa, while the inhabitants of the
island were perfectly healthy.   As soon as the nature of the
disease was known, they put to sea, in hopes of receiving advantage from a pure circulation of air ; but the sick-list increased
daily, and they returned to the harbor of Curaçoa in a much
worse condition than they left it.   Forty sick were landed ; and,
though there was no restriction in intercourse with the inhabitants, there was not a single suspicion of contagious influence."

Dr. Nathaniel Potter quotes the remarkable case of the Busbridge Indiaman, which sailed from England to Madras and
Bengal on the 15th of April, 1792 ; on the 26th of May she
crossed the equator in 27° west longitude, i. e., passing far west
of the Cape de Verde Islands and near St. Paul's Islands, so as
to enter the yellow-fever zone.   The mercury ranged from 80°
to 96° ; and "in this state of things the yellow fever broke out,"
says Dr. Brice, "although we had touched at no port, nor had
any communication with any vessel."

On the 3d of June, 1799, the frigate General Greene left
Newport, R. I., for Havana, which she reached on the 4th
of July.   She was beset by a storm which lasted several days,
and she made a great deal of water, *notwithstanding that she
was a new ship. A great heat prevailed ; her ballast containing
earth charged with organic matter, and a quantity of cod-fish
on board, putrefied.*   Notwithstanding all the sanitary measures
taken immediately, yellow fever broke out.   It gradually ac-

1763, as so graphically described by Rouppe ; on board of the Majestic, at Port
Royal, Martinique, in 1795 ; of the Ganges ; of the Peacock and Grampus ; the Sea
Island at Middletown, in 1820 ; the Polly, at Chatham, Conn., in 1796 ; the Ten
Brothers, at Boston, in 1819 ; the Favorite and Ocean, at Perth Amboy, in 1811 ;
the various vessels from New Orleans, at Staten Island, in 1848 ; the Bann, at As-
cension, in 1823 ; the Palinure, at Martinique ; the Néréide ; the Expéditive ; the
Gloriole ; the Eglantine ; the Africaine ; the Middleburg ; the Chichester ; the Her-
minie ; the Vestal ; the La Ruse and the Grayhound, at the Wallabout, in 1804 ;
the Ann Maria, at the New York Quarantine, in 1821 ; the Alban, at Port Royal,
Jamaica ; the Snake, at St. Jago, Cuba ; the Tartarus, Crocodile, Dee, Satellite
Hecla, Megæra, and others, of which we have the records." (La Roche, vol. ii.,
p. 423.)

quired virulence and activity, until the ship touched the port of
Havana, where at the time the disease did not exist.   Dr. Holli-
day of this city identified the disease.   This ship suffered again
in 1800, the disease proving still more malignant and fatal, and
attacking some individuals from the town of Newport, who,
after her arrival, worked on board.   The air in the hold of this
vessel, when first inspected, was so contaminated as to extin-
guish lights immediately.

In September, 1828, when Vera Cruz was quite free from
yellow fever, the United States ship Hornet, lying three miles
off at Sacrificios, suffered from an outbreak.   The ship had been
repaired and " salted," was very damp, and foul with fetid bilge-
water, putrid wood, chips, and shavings; and the external temperature averaged 87° at noon, ranging from 79° to above 90°.
She left for the north, and as the thermometer descended to the
freezing point the disease entirely disappeared.

The United States ship Levant contracted the disease, owing to
the foul state of its hold, in 1841, on a cruise by the West Indies
to Pensacola, and it only disappeared after several severe frosts.

A Court of Inquiry, ordered by the Secretary of the American Navy, dealt with an outbreak of yellow fever on board the
frigate Macedonia, in the summer of 1822.   She sailed from
Boston for the West Indies on the 2d of April, reaching Havana
healthy at the end of the month.   Here she lay till the 4th of
June.   She subsequently touched Cape Haytien and Port-au-
Prince.   All these places were remarkably healthy.   Yellow
fever broke out on board on the 8th of May, extended rapidly
among the crew, and carried off several of the officers.   On the
14th of June she sailed for San Domingo, but the disease continued, and on the 24th of July she sailed from Havana for the
Chesapeake.   The officer in charge found at Havana that the
air in the ship's hold was dense, close, disagreeable, and hot.
Her ballast was found muddy and dirty.   Water was let into
the hold between the 28th of April and the 7th of May.   The
captain having been informed by some English officers, who
arrived there after the Macedonia, that there was a "standing
order in the English service that water should not be let into
their vessels in the port of Havana," the practice was discon-
tinued.   The bilge-water then became filthy and offensive, so

that the men were dispersed and sent off from the ship when it was pumped out. "There was a gelatinous substance of a very offensive character on the chain cable when hove in; and on taking out the starboard cable a part of it was found to be wet, in consequence of a leakage from one of the berth-deck scuppers. The casing of this being removed, about two bucketfuls of very offensive filth was found."

"The sickness and mortality on board, according to Captain Biddle, were greatest among the persons employed in the hold and among the carpenter's crew, who, by working the pumps, were most exposed to the bilge-water discharged from the ship; and by Dr. Chase, one of the surgeons, it is stated that the disease commenced near the pumps. The boat crews were, on examination, found to have suffered less than the rest of the ship's company. It appears that the awning was constantly spread while in Havana, and that the men were very little exposed to wet or to the sun, or to duty in boats, or to fatiguing duty on board, being excused from keeping watch at night.

"From these various circumstances the inference is natural that the disease, which carried off one hundred and one individuals out of a complement (including officers) of three hundred and seventy-six, arose from the operation of causes located in the vessel itself. There was, as we have seen, no yellow fever at the time in the city or port of the Havana, and the same fact has come to my knowledge through other channels. Other vessels, at no great distance from the frigate, did not suffer from the disease, and neither officers nor men could have communicated with individuals already affected. Hence, it was impossible for the fever to have arisen from the introduction of a contagious germ, or from morbid effluvia proceeding from the port and wafted from a distance. The disease thus produced continued on board during the passage to St. Domingo, and therefore could not have depended on a cause existing in the harbor or city of the Havana, for in that event it would have ceased soon after the vessel put to sea. It increased at Port-au-Prince; and, as the fever was not prevailing there at the time, its aggravation was not the result of external influences having their source at that place. It is to be remarked that the disease continued to prevail during the passage back to the Havana, as well as during the stay there, and did not cease before the arrival of the vessel at Norfolk and the landing of the crew. To this must be added that the ship's hold was the receptacle, while at the Havana, of materials which in other localities have, under atmospheric and thermometric conditions of like nature, given origin to morbific effluvia of a most pestiferous character; that the disease first attacked and prevailed most severely among those most exposed to the effluvia from the hold; that the boat's crew, who were less exposed to those effluvia or to the contaminated atmosphere of the ship, were less affected by the fever than those who remained on board; and that the medical officers of the ship, Drs. Cadle and Chase, and Dr. Dayers of the navy, whose opinion was asked by the court, and who was familiar with all the circumstances of the case, expressed the opinion that the fever originated from noxious effluvia generated in the ship's hold.*

"Scarcely less important," says La Roche, "than the preceding, is the history of yellow fever on board the United States brig Enterprise, in 1822. In the middle of May she was three days off the Morro Castle, Havana, which is believed, after careful investigation, to have been free from the fever in an epidemic or sporadic form. She sailed for Charleston near the end of June, and remained there eight days; but when she reached there one of the lieutenants was sick. He died on the 1st of July. The cases multiplied; the ship sailed for New York with ten on board, and the next day they increased to thirteen. On the 2d of August, twenty-five days after her arrival, and after repeated whitewashing, letting in water, and constant ventilation, one of the sailors obtained permission to take his wife on board; this woman was taken sick with yellow fever on the 9th of August, and she died in the Marine Hospital on the 18th of that month."

This ship Enterprise is referred to in a private letter to me from Dr. J. C. Faget. He remarks that Mélier would have called her "sick with the yellow fever" (malade de la fièvre jaune). She was the subject of some special observations by Dr. Carpenter, who stated that she had been often disinfected according to methods then known, such as aëration, chloride of

---

* La Roche, vol. ii., pp. 429, 430.

lime, etc.; but, whenever she sailed in the hot tropical seas, there were invariably sailors affected with yellow fever. But at last she spent a winter in New York and was cured. Thereupon Carpenter declared that only cold would destroy the germs of yellow fever.

"Other cases of *evident infection generated on board American vessels*, and spreading among the officers, crew, or passengers, or on land in the vicinity, are on record, and are of a nature which admits of no doubt." So states the learned compiler,* who proceeds with further references, which I venture to transcribe and abridge:

"In 1799 the sloop Mary was sent into Philadelphia, as a prize to the ship of war Ganges. She was not from a sickly port, and at the period of her arrival there was no one sick on board. As soon as her cargo was removed, her decks were washed, and the hatches and ports all shut down. In this way she remained three weeks, the weather being at the time very hot and dry. The hold and the interstices between the timbers contained a quantity of vegetable matter (coffee), which, being mixed with the bilge water and that which had fallen from the deck at the time of washing, aided by the high temperature and close confinement of the air, fermented, and gave rise to the development of highly offensive effluvia. 'The noxious effluvia,' says Dr. Caldwell, 'that were generated in abundance, having no vent to escape and be dissipated in the atmosphere, mingled with the air in the vessel's hold, and produced it in an extreme degree of vitiation. A smell resembling that of common bilge-water, but more offensive, became troublesome to those engaged about the wharf, and was at length traced to the place where the Mary lay. She was soon suspected as the source of the nuisance. Her ports and hatches were accordingly thrown open, when the foul air rushed out in torrents, and spread through the neighborhood a suffocating stench. Several persons exposed to these effluvia were a few days after seized with decided symptoms of yellow fever.'"

In 1803 the Hibbert, a British three-decker, left Portsmouth, England, in ballast, and arrived in New York on the 3d of July, where she had orders to load with pine, for the bay of

* La Roche, op., cit. vol. ii., p. 432.

Honduras. The workmen engaged in loading the ship found that the ballast, composed of sand, had not been changed for many years, and that the woodwork as well as the bridges were covered with excreta. The stench occasioned by the collection and removal of this offal was intolerable. The men were obliged to run to the port-holes and hatches for fresh air. They removed all centers of putrefaction, but several of the workmen suffered from hæmorrhages, black vomit, *yellow fever*, and some died very rapidly. No one could imagine that a ship which had left England could bear the morbid principles in so violent a form. Nevertheless, after strict inquiry, it was proved that the Hibbert, employed in 1801 to transport soldiers from Portsmouth to Halifax, had from thence served to transport another regiment from Nassau to the Bahama Islands, and from thence she returned to Portsmouth with a third regiment; that in this triple voyage, as well as in that she made from Portsmouth to New York, *the same ballast had been in the ship, and she had been totally exempt from disease.* After leaving New York, the Hibbert, incompletely disinfected, lost several sailors on her passage to Honduras, and there she was the cause of the deaths of several persons, but, adds Devèze, " only of those who went on to the ship and there became infected."

In 1821 yellow fever broke out in the tropics on board the English armed transport Dasher, the frigate Pyramus, and an unarmed transport. " The Dasher left Barbadoes for Tobago on the 26th of August, but, in consequence of severe gales, was obliged to go to St. Lucia. Proceeding thence to Tobago, she there received on board a company of the Ninth regiment, and sailed for Grenada, in order that the men might avoid the endemic fever of the former place. This company, while on board, was perfectly healthy. On their landing at Grenada, the men were immediately placed in quarantine, and remained so for the space of fourteen days, during which period not an individual sickened. ' The Dasher, after landing this company at Grenada, proceeded hither (Antigua); but, a few days before she reached this port, yellow fever made its appearance among her crew, and previous to her arrival six men had been attacked, two of whom died.' The crew was landed, and the disease ceased among them. Blacks were employed to remove the ballast and clear

the hull. At the urgent solicitation of Dr. Hartle, the limber-
boards were taken up."

Here lay the mischief. Carpenters' shavings in great quantity were discovered in a state of putridity, choking up the lim-
ber-holes so that the water could not pass to the well of the
pump, and lay stagnant. Dr. Hartle declared that the fever
"was generated on board, by noxious effluvia received into
crowded and badly ventilated berths"; for, the moment the
crew and marines were removed, the disease ceased. Nothing
like the most distant appearance of contagion could be traced,
for none but those residing on board the ship, or exposed to the
effluvium from her hold, prior to her expurgation, suffered by
the fever.

"The Pyramus left English Harbor, Antigua, perfectly
healthy on the 19th of October for St. Kitts, where she remained until the 28th, when she sailed for French Harbor. A
day or two prior to her arrival at the latter place, fever of a
most alarming type made its appearance among the officers and
crew. The sick were landed, and the ship sailed for Barbadoes;
but the disease continued to prevail. A medical board was assembled, in order to investigate the probable cause of the sickness. The vessel, at the suggestion of the board, sailed from
Barbadoes, and cruised as far as 28° north; but finding this
avail nothing, and that the disease became more alarming, the
captain hastened to English Harbor, where he arrived on the
3d of January, 1822. The crew was there landed, and the ship
emptied of her stores, shot, tanks, ballast, etc. On the opening
of her hold, the effluvia which issued surpassed anything Dr.
Hartle had ever witnessed, and affected every one exposed to
its influence.

"The quantity of filth which was taken out was sufficient to
fill four large mud-boats, consisting of shavings mixed with coal-
tar and the water which, in consequence of the choking of the
pumps, had accumulated under the limber-planks. All the
cases which occurred during the process of expurgation were
easily traced to exposure to this bog; and Dr. Hartle very justly
refuses to refer the disease to the influence of English Harbor,
inasmuch as other ships of the squadron that lay much longer
there escaped the infection. It may not be amiss to remark

that the sick of this ship were landed and placed in Antigua dockyard on the 15th of January; that between that day and the 30th only eighteen cases occurred; but that on the 31st six fresh attacks were added to the list, and the disease again appeared with increased violence and malignity. Dr. Hartle adds: 'This sudden reappearance and violence of the disease induced me to believe that the people had some communication with the ship, which was then undergoing a general expurgation. This, with a little trouble, I ascertained to be the case.' Changes were made in the distribution of the convalescents and the rest of the crew, and the disease was put a stop to completely. The crew reëmbarked on the 14th of March in excellent health, and remained so.

"As regards the transport above alluded to, the disease broke out on her passage from Barbadoes to English Harbor. The sick and all the troops on board were landed, and the vessel, after a partial cleansing, proceeded to St. Kitts; whence, having landed the stores and baggage, she returned to English Harbor. She then underwent a general purification, when a portion of the troops reëmbarked and sailed for Grenada, where they arrived in good health, the fever not having reappeared on board.

"It is a pleasing reflection," adds Dr. Hartle, "and a source of great gratification to me, that, notwithstanding one hundred and forty-seven cases of yellow fever, as distressing and malignant as any I before witnessed, have been by the three vessels imported into this island since September, 1821, we have not had a single instance of any individual but those directly exposed to the local causes being attacked."

As we peruse these cases, and remember that decayed timbers and rotting wood-shavings have been frequently charged with mischief, a hint may be given to future observers carefully to look for the connection between the decomposition of different organic substances in ships in the tropics and yellow fever.

"Dr. J. H. Dickson, in a report to the Transport Board, alludes to the generation of the yellow fever in several vessels, the Blonde, Gloire, Star, Wanderer, and particularly the Dart. The disease broke out on board of the latter in April, 1807, and was satisfactorily traced to effluvia exhaling from offensive matter collected at the bottom of the water tanks.

4

"'So many people,' Dr. Dickson remarks, 'were taken ill after going on board this vessel, lying guardship at Barbadoes, that it was difficult to account for it, except on the principle of contagion, until the peculiar construction of the ship—viz., her being divided into compartments below, so as to prevent the circulation of air, and the stowage of the water in *bulk*—was . adverted to; and, on examining the large tanks or cisterns, their bottoms were found covered with an offensive slimy mud or deposition.' On the removal of some of these causes, the knocking down of the bulkheads, and the cleansing out of the cisterns, the fever was put a stop to. In the Thetis, in 1809, 'the fever did not appear until the hold was broken up, when about a dozen of men so employed were taken sick, and four out of five carpenters who lifted the limber-boards died.'

"In his account of the fever which prevailed on board the Nyaden, the surgeon observes : 'In clearing the after-hold, which was very offensive, several men immediately took the fever, some of whom died. This effect,' continues Dr. Dickson, 'is the more evident when contrasted with the healthiness of some vessels *close* to them, which were either accustomed to the climate or differently employed.' Well could Dr. Dickson, with these facts before him, remark : ' *The power of impure but strictly local effluvia in producing yellow fever on board ships also, as well as on shore, is unquestionable.'* "

Dr. Gillespie * speaks of the events on board the ships cruising in the West Indies during the years 1795–'98, and remarks that at that time all the vessels of the squadron had their crews in good health, except the frigate La Pique, which had been captured and carried into an English harbor, Antigua, to be refitted.

"In the beginning of November that frigate arrived at Martinico, and the remains of the crew had acquired a good state of health, though they had the sallow complexion which men generally have when confined in impure air. November the 12th she sailed for Barbadoes, having received a draft of seventy-five men from the Ganges. From being embarrassed with a convoy, and from unsettled southerly weather, the passage was long; two hundred French negroes were taken out of a vessel, which

* " Diseases of the Leeward Islands Station."

was in danger of foundering, and were kept on board the Pique until her arrival at Barbadoes. They were confined some time in the hold. Such a mixture of men, strangers to each other, has been often found to occasion sickness in ships, and, together with other causes, fatally operated here before the arrival of the ship at Barbadoes. A malignant yellow fever had made its appearance, and continued to rage with destructive violence among the crew of the Pique, and is supposed to have proved fatal to a hundred and fifty men. Out of the Ganges draft, twenty-eight alone are said to have survived the epidemic. The negroes, it is probable, were saved by being disembarked on the arrival of the Pique at Barbadoes. This," continues Dr. Gillespie, " is a melancholy instance of the generation of a fatal epidemic on board of a ship, at a time when the inhabitants of Barbadoes and the crews of the other ships in company remained free from any such disease."

La Roche says that Dr. Gillespie being, as every page of his volume attests, fully competent to discriminate between true yellow fever and other forms of febrile disease observed on shipboard, his statement of the occurrence in the ill-fated vessel " will of itself be sufficient *to scatter to the winds all the doubts entertained by fanatic contagionists as to the possibility of that generation.*"

Dr. Trotter publishes the following interesting letter from Dr. Crawford, relating to Port-au-Prince, in his " Medica Nautica ": " On the capture of the port, June 4, 1794, about forty sail of merchantmen were found in the harbor, most of them large vessels, the cargoes of which were coffee, cotton, sugar, and indigo, that had been stored in them from one year to three, in which time many of them never had their holds opened, from the disturbances that prevailed among themselves. On board of them men were sent from the whole squadron, to fit them for the passage to Jamaica, which, from the state they were in, was not to be soon done. The weather was excessively warm, and some days elapsed without a breath of wind. The very first day the people proceeded to work, many were taken ill, and sent on board their respective ships with fever; several with strong convulsions that were succeeded by fever, and one or two died. I was sent to several to remove the sick, where I found the stench from the damaged coffee and sugar almost insupportable; it wanted no degree of penetration to prognosticate the consequences in two, where there was a quantity of sugar, etc., melted in the hold. I saw the mixture in an actual state of effervescence, and bubbling up from every part. From these ships, I can vouch, the disease was first introduced to the Penelope. I most truly think that the primary cause of this horrid disease originated from these ships. One thing is most certain, that, if it did not originate there, it was much increased in violence by our connection with them. After they were fitted out, on their passage to Jamaica, the lost were then three to one in comparison with the men-of-war, although this passage was not more than seven days. In the Horizon, on board of which were Lieutenant Gaeren and Mr. Stupart of the Europa, the crew had been replaced three times, and from illness got in her died thirty men. Seven out of ten died on the passage to Jamaica; another of them was picked up at sea by a Guineaman, with every soul dead on board; even a number of negroes, who afterward cleaned them, died from fevers caught on board of them."

In 1800, at Demerara, where, according to Drs. Ord and Durkin, no fever existed, "a ship arrived about the beginning of July or end of June from Liverpool, after touching at Surinam. The filth on board, occasioned by a cargo of horses and the extreme neglect of the officers and crew, was such as beggars description. Infection was the consequence. Her officers were the first sufferers; every man died. All who went on board were attacked, within thirty hours after, with a fever of infection. What a lesson this to masters of vessels! How clearly it exhibits the necessity of exertion on their part to maintain cleanliness on board their ships! And how evidently does it display their responsibility to the public for the consequences of misconduct!"

La Roche declares this to be as clear a case of the generation of the yellow-fever poison on shipboard as could be desired. It is indeed admitted as such by Dr. Chisholm, the stern advocate of contagion.

Chisholm cites another highly interesting case, showing that fermenting wine may be as bad as decayed wood or stinking molasses. "Ships containing wine in their holds in a state of decomposition are generally extremely sickly, and the character of the prevalent disease is that of yellow remittent fever. Several instances of this took place in Fort Royal Bay, in the years 1797, 1798; and the situation of the ships in the open bay, far from the influence of marsh effluvia, precludes a suspicion of the fever proceeding from that cause. The ship Nancy, Captain Needs, from Fayal, with a cargo of wine for the army, arrived at Fort Royal, Martinico, in the month of October, 1798. She met with a gale of wind at sea on the 17th of September, and several of the casks, from the motion of the ship, became leaky. The captain was actually taken sick at sea, and died with every symptom of the highest grade of yellow remittent fever. The mate and several of the crew were attacked with the same complaint; they recovered; but a mate shipped at Fort Royal fell ill on board and died. The ship lay out in the open bay; no vessel near her was sickly; and she herself became very healthy after the cargo was landed."

The Regalia sailed with black recruits from the coast of Africa for the West Indies in 1815. The ship was good and crew healthy until she took on board on the African coast a quantity of green firewood. The ballast had never been changed or shifted from the time the vessel left England. It was shingle ballast, with a considerable mixture of mud and other impurities. The Regalia was far from a dry ship, and the absorption of sea-water among foul ballast and green wood could scarcely fail to prove unwholesome. Yellow fever broke out on the voyage, and continued up to the time of her arrival at Barbadoes. The blacks were exempt, and though the Regalia communicated freely with the seaports of Barbadoes and other islands, landing the sick or dying subjects of that disease among the inhabitants or at the hospitals, the infection was not communicated anywhere.

I shall not quote the histories of outbreaks on the Rattlesnake in 1824, or on the Éclair in 1845,* which have been

* " Were it necessary, similar instances of this mode of origin might be adduced from the history of British vessels—the Childers, the Isis, the Ferret, the Scylla, the Thracian, the Iphigenie, the Wasp, the Tribune, the Farmer, the Bustard, the Pylades, the Antelope, the Tigris, the Scamander, the Blossom, the Kent, the Circe, the Trinidad, the Serpent, the Brazen, the Busbridge, the Pompey, the Bedford, the

viewed by some as typical of original ship infection; but I shall follow La Roche in his search among French writers. I make no apology for entering into lengthened details on this evidence, since they are essential to an understanding of the question.

Dalmas relates an outbreak of fever on the Souverain, a seventy-gun ship, on her passage from Europe to the West Indies. Chervin gives several examples of yellow fever at sea before the infected vessels had reached West Indian ports.

Yellow fever broke out in the brig Fabricius of Marseilles in 1818, on her passage to Fort Royal; and, before reaching any point where infection might be due to the wafting of the effluvia from the shore, yellow fever broke out.

"The French brig of war the Euryale, commanded by M. Villaret de Joyeuse, was attacked with the yellow fever while on a cruise, and compelled in consequence to seek shelter in Fort Royal, Martinique, about the close of June, 1821. Before reaching that port the Euryale had already lost six men, and among them the surgeon; and at the time of her arrival the sick-list was very large. The sick were sent to the hospital, and the convalescents removed to Fort Bourbon. In neither of these places was the disease communicated to the attendants or others; but, on the other hand, a number of men who were sent to work on board were seized with the fever, and several died. In this, as in other instances, the disease did not extend beyond the focus of infection where it had originated, and where it affected those who exposed themselves to its action.

"Dr. Lefort, to whom we are indebted for the above case, adduces, as additional proof of the local origin of the yellow fever on shipboard, the account of three other vessels of war. The Egérie, the Diligente, and the Silène, which, during the sickly season of 1821, anchored at Trois Islets, a port situated at the bottom of the Bay of Fort Royal, Martinique. During their stay, the Egérie was attacked with fever and lost a great many men. She was ordered to sea; but, the disease increasing, instead of being arrested, she reëntered the port, and there went

Powerful, the Pilot, the Scout, the Volage, the Vestal, the Skipjack, the Rainbow, the Magnificent, and others; but enough has already been said on the subject to place the question of that origin beyond the possibility of a doubt." (La Roche, vol. ii., p. 447.)

through the usual process of purification. While the Egéric was at anchor at the Trois Islets, the intercourse between her and the Diligente and Silène was in no way prevented ; notwithstanding which, these vessels remained perfectly free from the disease. On the 19th of October the Diligente proceeded to Fort Royal, and soon after was itself attacked with the fever. On the 30th she was ordered to sea in company with the Silène, and during the passage from Martinique to Porto Cabello, but more especially after a week's sojourn in that place, suffered extensively from the disease. During the whole of this time the Silène continued free from the infection, although daily visited by men from the suffering vessel. It is scarcely necessary to remark that if the Egéric had derived the disease from the atmosphere of the Trois Islets, or from contagion, the Diligente and the Silène would, being exposed to the same causes, have shared a similar fate ; and, had the Diligente not suffered from the action of some cause inherent to the vessel itself, it is difficult to understand how its companion, the Silène, could have escaped so effectually the inroads of the fever, exposed as it was to the same atmosphere, communicating in the same way with the shore, and visited frequently by the crew of its infected companion." *

Type of Disease in the Eastern and Western Atlantic.

The following case suggests the idea that a foul ship, being sickly in the Eastern Atlantic, may gradually develop, as the western shores are approached, a genuine yellow fever. I have omitted mention of many cases (they are literally innumerable) · of the disease breaking out some days after leaving the West Indies. My main object is to insist on close investigation in a comparatively new direction ; and it is a matter of first importance if we can determine, as I feel sure we can, that yellow fever is in its essence not a shore disease, but rather creeps on shore from infected vessels, and is the product of putrefaction and ordinary fermentation, under conditions similar to those which would develop hospital gangrene, puerperal fever, and pyæmia.

"In the year 1802 a flotilla filled with French troops sailed from Tarentum for the island of St. Domingo. The vessels

* La Roche, vol. ii., pp. 448, 449.

consisted of small Neapolitan polaccas, under the escort of a frigate, each of which, though only intended for the accommodation of at most one hundred men, received one hundred and fifty. Encountering, soon after leaving the port, a severe storm, the vessels were dispersed, and sought shelter where best they could. They reassembled at Leghorn, and thence proceeded to Cadiz, there to join another division of troops that were to form part of the expedition. Stopping again at Cartagena to take in proper provisions and to refit, the expedition set sail for St. Domingo—the troops being now transferred to eight vessels freighted for that purpose. Of these vessels, one was set apart for the accommodation of the sick.

"The spring had been cold and wet. Summer came on suddenly, and was characterized, during the months of June, July, and August, by intense heat. Soon after the departure of the vessels from Cartagena fever broke out on board, and continued to prevail in some of the ships till their arrival at Cape Haytien —spreading more extensively and acquiring greater malignancy as they approached the tropics and were exposed to a higher temperature. The disease, without doubt, consisted of one of the forms of true yellow fever; but it exhibited, especially at first, a mild character. At the time of arrival, the yellow fever, in all its purity, was prevailing among the troops at St. Domingo ; and, by comparing the symptoms presented by the cases on board with those noticed ashore, the surgeon of the squadron was enabled to convince himself of the identity of the two diseases.

"In this instance there can not be the remotest reason for referring the disease to any other than a local origin. Nothing is said of the soldiers having imbibed the cause at Leghorn, Cadiz, or Cartagena, in neither of which places, indeed, it existed that year. Dr. Bégueric lays some stress on the effects of bad regimen, but especially on the excessive heat to which the men were exposed; as also on vicissitudes of temperature and exposure to night air; but he likewise attaches much importance to the morbid exhalations arising from the accumulation of the troops and the decomposition of animal and vegetable substances contained in the vessels. It may not be improper to add that the disease manifested no contagious property." *

* La Roche, vol. ii., pp. 451, 452.

Contagionists on Spontaneous Origin in Ships.

Contagionists have certainly denied, and denied emphatically, the production of yellow fever anywhere on land. They admit the possibility of its origin in ships within narrow bounds, and perhaps with justice, "rejecting as unfounded and even absurd the idea of including within these any section of the temperate zone." . . . "The facts," continues La Roche, "presented in the preceding pages, and which have been collected from sources entitled to full confidence, are of a nature to overcome effectually all the objections raised against the reality of the generation of the yellow-fever poison on shipboard."

"The facts that have been adduced," he adds further on, "establish, beyond the possibility of a denial, the reality of the development of the yellow fever on shipboard, from the operation of causes existing therein, and unconnected with any contagious or infectious germs introduced from without."

A case is recorded by Dr. Rancé. In 1852 the bark Flora arrived from Bordeaux at New Orleans. She had necessarily to touch the tropical zone of the Atlantic. She took on board some perishable goods, and, these becoming putrefied toward the end of August, some cases of yellow fever appeared in her in mid-ocean.

These cases suggest the possibility of an outbreak of yellow fever occurring, as on the ship Plymouth, when a tempest overtakes a vessel, and the provisions for ventilation, always defective, with hatches battened down, are rendered impossible. A mass of human beings contaminating a hot and humid atmosphere, putrefaction actively setting in wherever decaying timbers, organic matter, heat and moisture, and unavoidable exclusion of light, are found combine to constitute conditions which, in a certain part of the tropical Atlantic zone, may result in yellow fever. The Plymouth had been infected before, and this leads to the belief that the germs of yellow fever remained in her; whereas in all probability she is a rotten ship, and will be foul and unsafe for all time in the West Indian seas.

Dr. E. M. Schaeffer, of Washington, has kindly directed my attention to some important contributions on yellow fever in the "Army Statistical Reports of Sickness" from January, 1855,

to January, 1860. Assistant Surgeon La Fayette Guild, report-
ing on the disease at Fort Columbus, N. Y., in 1856, says:
" Yellow fever, so long the scourge of the tropics, has for the
last few years been steadily and progressively encroaching upon
the temperate regions of our country." His observations were
confined to Governor's Island, in latitude 40° 42′, longitude 74°
9′, situated in New York Harbor, at the junction of the North
and East Rivers with the bay, and about 22 feet above low-water
mark. He says: " All agree that yellow fever can be generated
on board of vessels. Whether the morbific agency depend upon
the decomposing timbers, the bilge-water, the putrefying cargo,
or what not, still the fact is beyond a doubt that the disease can
be so produced. The first case that occurred in this harbor last
summer was on board of a vessel, and not only did one vessel
arrive in this port with the disease on board, but several; and
they were all collected and placed in quarantine, whence, in my
opinion, the disease was disseminated throughout the different
places along the bay that were not protected from the wind that
swept over this congregation of infected ships, filling the air
with their pestilence. The facts that substantiate this opinion ·
are that the epidemic broke out simultaneously in different localities on shore, after the ships had been placed and the process
of unloading and cleaning them had been commenced ; that the
disease was confined to the southern portion of Governor's Isl-
and, and to those habitations which were openly exposed to the
winds from the quarantine ; that the intervention of their board
partitions, trees, shrubbery, and elevated portions of the ground,
which served to shut off these winds, gave to individuals so protected an immunity, and that the removal of the inmates of
South Battery to a place where they were protected from these
winds checked the ravages of the disease among them." (Pp.
18, 19.)

Further on he says: " It certainly is a question that should
engage the studious attention of boards of health, and of those
members of the medical profession whose duty it is to regulate
the quarantine enactments and to guard the public health against
the pestiferous influences of infected ships that annually arrive
in our cities on the seaboard." (P. 23.)

Surgeon Charles McDougall reports on an outbreak at Fort

McHenry, Maryland, from the 16th to the 26th of September. I quote it for the definite information as to the health of the fort at the time of the outbreak, and the probability of the impure air of the ships being wafted on land with sufficient virulence to propagate the malady.   He says: "After a careful and critical examination of Fort McHenry and its immediate localities, nothing could be found which would predispose to or engender disease.   The police was perfect; the drainage as complete as could be desired ; no standing water ; clothing and diet of the soldiers excellent; the general habits of the command perhaps better than at most of our military posts.   How, then, could the yellow fever originate, or whence contracted ?   All of the cases but one occurred in the row of frame shanties northeast from the fort, some eighty yards.   Five of the inmates were attacked by the fever about one time, and all within a period of eight days.   The shanties were clean and well whitewashed, but old and deficient in sub-ventilation.   We can imagine that the sweepings and washings of years might have produced a condition like that of defective sewerage, which would favor or induce disease.   But most unlikely in regard to those shanties.   We believe that the yellow fever at Fort McHenry came from vessels infected with the disease at quarantine, near by ; that the infecting poison was carried by the winds to the frame shanties facing the bay ; and those who inhaled the poison, being in condition, took the disease.   The command in the fort were protected by the ramparts ; not one of them sickened. Why their immunity from an attack ?   There seems to me not a doubt that the *materia morbi* of yellow fever is *portable*, and where it finds favoring conditions will reproduce itself and extend its dreadful ravages." (P. 85.)

Surgeon B. M. Byrne, in a report on yellow fever at Fort Moultrie, speaks of the high reputation of the island for salubrity, but the disease entered it from Charleston with four thousand refugees ; it was communicated to twenty soldiers, of whom ten died and ten recovered.   The first case which had occurred at Charleston in July was in the person of a man who left an *infected vessel* lying at quarantine.

Assistant Surgeon A. F. Watson relates the invasion of Fort Brown, Texas, by yellow fever, and again shows how, when a ship has carried the disease, it may at times be propagated under the most favorable conditions. He says: "Previous to the breaking out of yellow fever the city and fort were remarkably healthy, and in the city there had been many improvements carried on for the last two years. The fort also was at the time in remarkably good and clean order. There was no apparent local cause of fever here. The meteorological register did not indicate any extraordinary condition of the atmosphere." (P. 183.)

Conditions essential to Yellow-Fever Development.

Dr. Faget considers that the three conditions *alone* essential for the development of yellow fever are:

1. Organic matters in a state of decomposition.
2. A maritime center—in fact, a ship—in the Atlantic.
3. A tropical temperature.

The weight of evidence coming to his support is great, and it carries conviction with it that it is the shipping in the tropical Atlantic zone which first becomes infected.

It has been incontestably shown that the malady exists in a region where a high degree of humidity as well as high temperature, still air, and ill-ventilated ships always coexist. These conditions alone suffice to lower the stamina and induce disease among picked seamen. No better soil could be found than in one of these vessels for the development of yellow-fever poison, should any specific germs have to be deposited for that purpose. But admitting for argument's sake their indispensable presence, is it a far-fetched view to take of the matter that the richly charged waters of the ocean in the tropics bear the elements essential to the development of this disease, by favoring a virulent putridity alone sufficient to generate that element? I have very carefully consulted with travelers, and all declare that in no city of the West Indies, of Central or South America, is yellow fever considered indigenous by the best informed. In this way, so far as the cities are concerned, the disease exists nowhere. The conditions of heat, humidity, and decay occur of course where paludal fevers (the Chagres and other fevers) abound; but here is one circumstance, and an all-important one, viz. : that ships, the partially submerged abodes of man, usually without any, or with very imperfect, provisions for ventilation, surrounded by hot and humid air above, and the warm waters of the tropics below, constitute wells of foul organic deposit in which there is an aggravation of conditions inimical to human life, well calculated to produce what Audouard called *nautical typhus*, and what in the West Indian seas is *yellow fever*. Closset in 1877 recognized its association, at all events, with *putrid typhus*, and so did Drogart.

Yellow Fever a Hybrid Typhus.

Yellow fever has been regarded as a hybrid form of typhus " by some physicians of a former generation—Blane, Lemprière, Dickson, and Chisholm." (La Roche.) It may be typhus modified by a special factor. Chisholm thought it was " the typhus of Europe grafted on the yellow remittent fever of the torrid zone."

It is of course a matter of the highest and most pressing importance to determine the identity or dissimilarity of the *vomito prieto*, or malarial black vomit·of the Mexican coast, and that of the lowlands of Louisiana. Are all cases of black vomit, so fatal in Vera Cruz, cases of yellow fever? Speaking of Mexico, Prescott says :* " All along the Atlantic the country is bordered by a broad track, called the *tierra caliente*, or hot region, which has the usual high temperature of equinoctial lands. Parched and sandy plains are intermingled with others of exuberant fertility, almost impervious from thickets of aromatic shrubs and wild flowers, in the midst of which tower up trees of that magnificent growth which is found only within the tropics. In this wilderness of sweets lurks the fatal malaria, engendered probably by the decomposition of rank vegetable substances in a hot and humid soil. The season of the bilious fever—*vomito*, as it is called—which scourges these coasts, continues from the spring to the autumnal equinox, when it is checked by the cold winds that descend from Hudson's Bay." Further on (p. 300, vol. i.), speaking of the sand-hill district of Vera Cruz, he asserts that " the bilious disorders, now the terrible scourge of the *tierra caliente*, were little known before the conquest. The seeds of the poison seem to have been scattered

* " The Conquest of Mexico," vol. i., p. 6.

by the hand of civilization; for it is only necessary to settle a town, and draw together a busy European population, in order to call out the malignity of the venom which had before lurked innoxious in the atmosphere." The periodical changes of site of La Villa Rica of Cortes, Old Vera Cruz and New Vera Cruz, have all been ascribed to the *vomito* afflicting the inhabitants; but, as Prescott states, they have gained little by the exchange. The mere fact of its always having been a seaport would suffice to keep up a true yellow fever; but, according to Faget, the city limits contribute the boundary beyond which it can not pass into the open country. Might the site of the city not have been changed because the disease did not invade the part chosen, as an experiment for future habitation, but that this immunity was due to " open space " ? The connection between *vomito* and European invasions is more than a tradition. No mention is made of the disease in the records of the conquerors. " The fact doubtless corroborates," says Prescott, " the theory of those who postpone the appearance of the yellow fever till long after the occupation of the country by the whites. It proves at least that, if existing before, it must have been in a very much mitigated form." *

Dr. Faget has no doubt whatever of the complete independence (*individualité morbide*) of yellow fever, and speaks of it as quite as distinct from other diseases as cholera or the plague. In this he is supported, as we have elsewhere shown, by the ablest authorities, old and new, and to him is due the credit of discovering distinctive signs, to be described in a later chapter.

Propagation of Jail and other Fevers.

Let us not forget the conditions under which some local, endemic, or indigenous diseases became virulent plagues.

Dr. Bancroft, in his " Essay on Yellow Fever," published in London in 1811, referred to the transmissibility of diseases induced at first by foul atmosphere, rendered impure by putrifying organic matter; and he says: " Many writers of celebrity, and among them the great Lord Bacon, have thought that no effluvia were so infectious and pernicious to mankind as those which issued from putrefying human bodies; and although a

* Op. cit., p. 394.

century and a half has elapsed since Diemerbroeck (' Tractato de Peste,' lib. i., cap. viii., p. 41) attempted to convince physicians that, at least, such effluvia could not produce the plague, yet the old opinion has kept its ground ; and it is still believed that, in their milder state, they must cause putrid fevers, and in their more concentrated state a true pestilence."

Bancroft had reason to believe that the yellow fever had been much more frequently produced than is generally imagined by miasmata resulting from the decomposition of vegetable and other matters in the holds of ships.

The late Dr. Murchison, of London, whose recent premature death is so deeply lamented by all who knew him, writing concerning the " Independent Origin of Typhus," says: " The conditions under which the poison is developed *de novo* are overcrowding of squalid human beings and deficient ventilation ; in other words, the poison is generated by the concentration of the exhalations from living beings, whose bodies and clothing are in a state of great filth. The intimate connection between the prevalence of typhus and overcrowding has been already demonstrated, and is generally admitted." Dr. Murchison directs special attention to the indispensable high temperature for the development of yellow fever, and says that it is quite possible that the alleged exemption of the Laplanders and Esquimaux from typhus, if true, notwithstanding the bad ventilation of their dwellings, may be due to the cold climate.

Typhus-fever poison has been supposed to be a chemical compound of ammonia ; but the particles of degraded animal matter, capable like pus-corpuscles of multiplying in a suitable soil, are probably the offending agents, according to Murchison. It is evident he had no idea of the compound exhalations, smelling of rotten eggs, in the tropical Atlantic, and which engender yellow fever.

Sir Gilbert Blane, in a letter to Rufus King, American Minister to England, said in 1799 : " It appears to me that the yellow fever can not be produced but in a season or climate in which the heat of the atmosphere is pretty uniformly, for a length of time, above the eightieth degree of Fahrenheit's thermometer ; that, under the influence of this heat, Europeans newly arrived, and more especially in circumstances of intemperance or fatigue in the sun, may be subject to it in many instances; but it has usually become general only by the previous influence of that infection which produces the jail, hospital, or ship fever, or from the influence of putrid exhalations."

Space forbids any lengthened reference to the contagion of the Egyptian plague, or to the putrid fever which attacked the survivors of the Black Hole of Calcutta imprisonment.

The reports of virulent infection in the form of jail and ship fever, when no cause but overcrowding and bad ventilation existed, are numerous and well authenticated. What Chervin did for the abolition of oppressive quarantine was accomplished to a much higher degree by the great philanthropist Howard, who devoted his life to prison reform. Prison typhus was common, and often passed beyond the prison walls. One of the most notable instances in history is that of the black assizes at Oxford in July, 1577, when the jail fever spread from the prisoners to the court, and within two days had killed the judge, the sheriff, several justices of the peace, most of the jury, and a great mass of the audience, and weeks afterward spread among the people of the town. Dr. Arnott remarks, " This was a fever which did its work as quickly as the cholera does now."

Pyæmia or Purulent Infection.

It is just over a quarter of a century since my brother, Mr. Samson Gamgee, of Birmingham, threw himself with great energy into the investigation of purulent infection in surgical wards, a subject then almost hidden from view. I pursued with him, and afterward alone, many experiments, and among other points ascertained that an animal might suffer little from the injection into the veins of an isolated dose of laudable undecomposed pus; but, if the pus were putrid, the infection was prompt and fatal. The history of contaminated hospital wards is full of interest in connection with the history of yellow fever in contaminated ships.

There is no natural nor artificial safeguard protecting the seaman in his tight dungeon amid the waters. Any leakage is not of pure air, such as actually finds its way through solid bricks and mortar, and which is the active preserver of human life in thousands of ill-ventilated dwellings. On land we are in a sea of air.   At sea the leakage is of warm and highly charged water teeming with organic matter, a rich mother-liquor for the most active' fermentation and rotting, so that a ship sealed, if possible hermetically, might burst under the pressure of accumulated gases. Read the case of the sloop Mary. She had been lying shut up, a war prize. Foul smells about the wharf (Philadelphia) directed attention to her condition. Her ports and hatches were thrown open, when the foul air rushed out in torrents, and the persons who inhaled it were seized with yellow fever.

Other Diseases propagated after Spontaneous Development.

America is signally free from all the purely contagious epizoötics, if we except the lung-plague imported from Europe, nearly fifty years since, into the dairies of Eastern cities. I have already alluded to the American cattle plague proper, originating in Texas, and which, like all indigenous maladies, when once understood is most easily prevented in its spread among northern animals. It is otherwise with the widespread and popularly designated hog cholera. This singular malady, attacking an omnivorous animal, when collected in large numbers from various sources on a limited area of ground, is developed like a virulent camp fever in man, and, strange to say, appears when once developed to be communicable to most warm-blooded animals, and I should say, under certain circumstances, undoubtedly to man. The varied developments of the disease are of striking interest in relation to those human disorders which spring from overcrowding, and in which nature's process of depuration produces at one time hæmorrhagic exudation, at another a malignant angina and catarrh, at another diarrhœa and enlargement of the intestinal glands, whence the disease has been confounded with human typhoid. The infected pigs die by thousands. They are purchased from healthy sources, and remain healthy where they are bred, but, whether or not confined in sties or the open air, the soil on which they lie and live becomes saturated with putrefactive emanations, which rapidly undermine the animals and breed in them a malignant and communicable disease. I am not attempting to be precise, and I only wish to show that there is no oddity, much less impossibil-

5

ity, in the manifestation of a disease from the congregation of living creatures under definite unwholesome conditions, which, once developed, spreads even by direct contagion.

A more familiar example in medicine is that of purulent ophthalmia of soldiers and children, which originates in over-crowded barracks and wards, and is rapidly propagated, causing intense suffering and frequent blindness, by the particles of de-composed pus, floating in the atmosphere, adhering to the towels and other objects, and often forced into the eyes in the innocent act of rubbing them, as a baby so often does with its little fist.

I might multiply instances of a spurious form of contagion which bears the same relation to a specific virus that the pus of soft chancre does to the discharge of a genuine Hunterian in-durated sore. It is as nearly as I can state in that relation that I consider yellow fever and an epidemic outbreak of small-pox in an unvaccinated community. The one is in its early outward signs scarcely distinguishable from the local marsh or paludal fevers—a hybrid typhus—and, until Dr. Faget's researches, scarcely capable of accurate and certain diagnosis. The other stamps itself from its invasion as an unmistakable eruptive fe-ver, due to a specific organic poison. The first spreads like mud pervading a pool by continuity from one disturbed center. The other dots the surface, and skips here and there in direct rela-tion to personal intercourse with the sick and susceptibility of the people. Both are deadly, but both vary materially in their course and development.

Difficulties of ventilating Ships.

Now, in relation to the ships which are such admirable pest-traps, I desire most forcibly to remind my readers of the great difficulty experienced in getting free currents of pure, dry air through the interior of a ship at sea. So essential is this that I hope the National Board of Health will investigate the subject and enact rules to be adopted by ports, favoring well-venti-lated vessels and condemning the stagnant tubs of pollution to protracted and prohibitory detentions.

The difficulty of clearing the hold of a deep vessel of foul air, without adequate blast, is much like the difficulty experi-enced in ventilating an old gasometer, on first entering which, after it has been left open for days and weeks, a workman, unless he be most careful, may promptly sacrifice his life and that of others who seek to rescue him.

Asphyxia in Ships.

Dr. David Boswell Reid, a worthy and able advocate in the field of hygiene, quotes in his treatise on the " Theory and Practice of Ventilation " (London, 1844) a narrative of the accident in the ship Minden in the year 1819–'20 (pp. 371–'72):

"On board H. M. Ship Minden, bearing the flag of Admiral Sir Richard King, in the harbor of Trincomalee, in the year 1819–'20, a boatswain's mate was ordered to see the pump-well swabbed out. A man was accordingly sent down with a bucket and swab ; but, as he neither filled the bucket nor answered when called to, a second man was sent down to see what he was about. He also refused to answer immediately. Three more rushed down into the well, who all, like the first two, remained silent. It was then reported on the quarter-deck that the men in the pump-well were supposed to have got into the spirit room. The master, on entering the cockpit, suspected the true cause of the men's silence, and ordered a lantern to be lowered into the well, the light in which went out when about half-way down. It was let down a second time, and the light burned long enough to show the whole of the men lying over each other on the keelson. On being lowered down a third time, the light was found to burn dimly at about six feet above the men. The master, with a bowling knot under his arms, descended the well, leaving directions to haul him up the moment he could not answer. He succeeded in slinging the men, who were hauled up and laid on the main deck, to all appearance quite dead. In a short time they began to respire; the lips and face became black ; they foamed at the mouth, and the whole were fearfully convulsed. None of them recovered their usual health, and some of them were invalided. The man who descended the well first was the first who recovered.

"(Signed) JOHN MILLER,
" Late Master of the Minden.

"London, December, 1842.

" Mr. Miller has since added :

"The officers who assembled in the cock-pit, on hearing what had taken place, remonstrated against the master descending into the well ; but he replied that he thought that he could breathe at least as low down as a light would burn, and on reaching the light he felt no inconvenience ; and even after he had succeeded in slinging the whole of the men, though he was engaged for a full quarter of an hour, he did not subsequently experience any further inconvenience than a slight headache. From the agitation, the air continued to improve gradually in the well, and, by the time he

had left, the candle was not extinguished when lowered to the keelson
(about six feet lower than it was capable of burning at first)."

It is easy to imagine how much worse must be the air in
many an old ship with rotten timbers and putrefying stores or
cargo in the tropics. How utterly impossible it must be to se-
cure health in such ships, and how easily a veritable ship-typhus
may actually originate there!

The Atlantic Yellow-Fever Center.

The Atlantic region, in which ships run special risk of yellow fever from contamination, is well known to present very
characteristic meteorological features, which tend to interfere
materially with the ventilation of ships. At the equatorial
calm-belt a conflict of winds occurs, for a northeast trade-wind
and a southeast trade-wind can not blow in the same place at
the same time. The air-particles have been put in motion by
the same power; they meet with equal force; and therefore at
their place of meeting they are arrested in their course. There
is consequently a calm-belt, as at Capricorn and Cancer.

In relation to the difference in the severity or character of
fevers engendered in foul ships crossing the Atlantic from east
to west, it must be remembered that the western half is far hotter than the eastern. High as the temperature is within 5° of
the equator north or south, the remarkable contrast between the
climatology of the sea and land must not be forgotten. Maury,
to whom I am indebted for my information on the climate of
the sea, remarks that "on the land February and August are
considered the coldest and the hottest months; but, to the inhabitants of the sea, the annual extremes of cold and heat occur
in the months of March and September. On the dry land, after
the winter is 'past and gone,' the solid parts of the earth continue to receive from the sun more heat in the day than they
radiate at night, consequently there is an accumulation of caloric,
which continues to increase until August. The summer is now
at its height; for, with the close of this month, the solid parts
of the earth's crust and the atmosphere above begin to dispense
with their heat faster than the rays of the sun can impart fresh
supplies, and consequently the climates which they regulate
grow cooler and cooler until the dead of winter again. But at

sea a different rule seems to prevail. Its waters are the storehouses in which the surplus heat of summer is stored away
against the severity of winter, and its waters continue to grow
warmer for a month after the weather on shore has begun to
get cool. This brings the highest temperature to the sea in
September, the lowest in March." *

Influence of Temperature of Sea-water on a Ship's Ventilation.

It is at first somewhat startling that the period of lowest sea-
water temperature at the equator should be about the period
when yellow fever gets its start northward; but we must here
consider the effect of temperature in the calm-belts on the ventilation of ships. Atmospheric pressure is high. Polar refrac-
tion is greater than the equatorial. The mean height of the
barometer in the calm-belts of the tropics is greater than it is in
any other latitude. There is an *accumulation* of air in the
tropical belt about the earth in each hemisphere. Could we
view our atmosphere from above, we should discover a ridge
and not a valley over the equatorial calm-belt. "In the belts
of low barometer—that is, in both the equatorial and polar
calms—the air is expanded, made light, and caused to ascend
chiefly by the latent heat that is liberated by the heavy precipi-
tation which takes place there. This causes the air which ascends
there to rise up and swell out far above the mean level of the
great aërial ocean. This intumescence at the equatorial calm-
belt has been estimated to be several miles above the general
level of the atmosphere. This calm-belt air, therefore, as it
boils up and flows off through the upper regions, north and
south, to the tropical calm-belts, does not so flow by reason of
any difference of barometric pressure, like that which causes
the surface winds to blow, but it so flows by reason of difference as to level." †

It is, I think, obvious that a ship in the calm-belts, and say
near the equator in March, is under the influence of the lowest external temperatures at its deepest part; it is likewise under the influence of a still, heavy, and saturated atmosphere

* "Physical Geography of the Sea," by M. F. Maury, LL. D., London, 1869,
p. 383.
    † Ibid., p. 355.

above. Every condition favors the gravitation of the foul air toward the bilge, and the only circulation is due to the rising from this of the gases of decay which are disengaged with great freedom. Air colder than the ocean-water would perflate its way through the ship, if any inlet could be found. A natural ventilation would be forced in spite of iron sides or sodden wooden walls. Hence, how soon a ship purifies itself on its northward passage, unless ports and hatches are fast! The trade-winds are the mighty purifiers of the Atlantic atmosphere within their reach. Insufficient they may be under overwhelming conditions of atmospheric impurity within ships, but their ventilating power is indubitable, and it is probably beyond their life-giving blast that those fatal influences are most rife which are capable of engendering yellow fever.

The Cargo.

Whatever may be the belief as to a ship in mid-ocean, under the requisite combination of circumstances, becoming infected with yellow fever without catching the disease in port, it is certain that, though atmospheric currents, even at sea, may with great difficulty bear on their course yellow-fever germs, the cargo can be of such a nature as to be either the carrier or the producer of such contamination as will engender the disease. A barrel of potatoes from an infected ship has been known to communicate the disease to a seaport town. This might and probably would occur by the adherence of foul ferments to the barrel or the potatoes. Sometimes, but rarely, a bale of cotton has been the *accused* carrier. Coffee is said to be usually innocent, and it is very generally acknowledged that the sound contents of an infected ship can be easily purified by free aëration. In the channels of commerce little harm has arisen from contaminated merchandise.

On the other hand, we have the most convincing testimony of the injurious and most probably sufficient cause of specific mischief in decomposing or decaying wood, wine, sugar, molasses, and bituminous coal. The fermentations started in the bilge-water by hydrocarbons, which drop or percolate into it, are certainly not innocent ; and the gaseous emanations, heated in the process of their production, permeate, to the detriment of human comfort and human life, into every part of the contaminated vessel. The sailors acquire a sallow and depressed look. Their exhalations add to the atmospheric filth. The saturated atmosphere prevents relief by the natural excretions, and, without regard to premonitory signs, without intelligent action to relieve the situation, a state of morbid indifference develops, which culminates in a pestilential outbreak.

Disinfectants not used at Sea.

I know from experience that it is the hardest possible thing to get skippers to use the most desirable disinfectants, which in my country must be carried by law. A few gallons of stinking carbolic acid—that curse of inefficient and blinding sanitation—of chloride of zinc or aluminium, could however, even if used, be of little or no service under such circumstances.

Frequently has attention been directed to contaminated ballast, and it is by no means a far-fetched supposition that ballast charged with putrifying organic matter may, in the tropics, foul to pestilential virulence the confined air of a ship. From this point of view the case of the ship Hibbert is interesting; for the only connection with the necessary focus in the tropical Atlantic basin was by ballast, which for years did no damage, infected no one, until ultimately saturated with putrefiable matter and subjected to intense heat. Then a disturbance of the filth, and probably that universal swabbing and washing which Dr. Turner exposes, established the link, with other favoring causes, to produce a true nautical typhus or yellow fever.

There are many instances of ships escaping infection in ports contaminated by yellow fever, and we have all heard of naval officers vaunting their conquests over the epidemic by cleanliness and free ventilation. The foul ships and the old ships are the yellow-fever ships; but the same would occur in relation to most diseases, and notably in relation to small-pox. A very exhaustive analysis of all cases, recorded or hereafter coming under observation, of yellow fever in ships, and of its uniform absence or frequent recurrence in certain ships, is imperatively demanded, and will more amply repay investigation than the study of yellow fever in towns. It is most dangerous to trust to tradition. Our knowledge, such as we possess, is based far too much on that already. The prompt investigation of last year's epidemic was most fortunate, and, although confined mainly to the infected cities, will undoubtedly contribute much to enrich our understanding on this question. It is well now that the National Board of Health is so closely watching the shipping. *There* is the mine for the yellow-fever pathologist; and, that once thoroughly explored, we shall wonder that, after man's experience in camp and jail fevers, we have been so remiss, so reckless, and so blind.

What relation does the cargo, in its nature, hold to outbreaks of yellow fever? Record, classify, and learn! Is it enough to have the exhalations from human beings in ill-ventilated ships· in the tropics to breed the disease? Who knows? Probably the animal matter from living creatures, with tropical organic decomposition and climatic conditions, complete the circuit for fatal invasions. To arrive at essentials, we must eliminate the non-essentials; and we know that in the North Atlantic, the Mediterranean, or the Indian Ocean, thousands of ill-ventilated ships, laden with fish or fruits and well filled with human beings, have been decimated—but not by yellow fever. The tropical something must be there from the West Equatorial Atlantic; hence yellow fever is indigenous, not to the soil, on which it never breeds, but to the ocean; there where, since ships traded, the disease has undoubtedly recurred, at intervals which seem to bear an inverse ratio to the activity of maritime intercourse. It is undoubtedly and incontrovertibly *indigenous* to ships, although it may be—nay, in all probability, is—*exotic* on every land. Its importation and persistence in certain harbors and cities are not more marvelous or inexplicable than the effects of *crowd-poisoning* in any latitude—the manifestation of the bubonic plague of old, and the dissemination of hog cholera.

As I am writing I receive the annexed extract through the friendly hand of a journalist, and, without committing myself to the view that Memphis received no fresh infection this season, I can not do better than constitute it the *finale* to this chapter.

" *Cause of the Outbreak in Memphis.*

" The true reason of the present outbreak of yellow fever lies not so much in the filthy streets and alleys of the city, but in the cupidity of some of our people, who would not give their consent to destroy even the bedclothes upon which patients died of the fever. It has never been demonstrated that the yellow-fever germ can be preserved through the frosts and breezes of the winter in the foul air of a privy vault; but it has been shown time and again that woolen goods, especially blankets, that have become saturated with the yellow-fever poison, will retain it for a very long period even in cold weather. It is well known that many persons in Memphis did not hesitate to preserve and even to sleep upon beds and bed-clothing that had been poisoned by the infected air of a sick-room, or by direct contact with the yellow-fever patient. These articles had been kept, of course, in bed-rooms where the heat of a fire during the day and the warmth of the sleeper's body at night prevented the germ from being frozen out. In many instances woolen clothing that had been hanging in the sick-room, where the air was reeking with the foul fumes of the fever, was packed away in trunks, or, with the poorer class, in wooden boxes. Here it remained during the winter. The warmth generated by the fabric was amply sufficient to preserve the germ in all its former vigor, and there it lay, like a deadly serpent, only waiting for the heat of summer to warm it into life. Mulbrandon's coat, which, like the shirt of Nessus, carried death in every fold, is now a matter of history. Another is that mentioned in the "Avalanche" of yesterday morning, of a South Memphis woman who has kept in a wooden box all the clothing of her husband, who perished by the fever last year, and even the bed-clothes on which he died, stained all over with black vomit. One of the ablest physicians in Memphis said not long ago that there was not a house in the city, whether occupied during the fever or not, that had not been thoroughly infected by the yellow-fever poison. It should be remembered also that even those who fled from the city when the fever broke out left behind them carpets, bedding, and winter clothes, to receive in trust for them the insidious poison which they were trying to escape. This reasoning may not be founded upon the principles of medical science, but it is certainly justified by common sense; and, by getting up all the evidence to be had in regard to the matter, the medical fraternity may be able to throw some light upon the origin of the present outbreak of yellow fever in our city.—(From the Memphis "Avalanche.")

How much like the dried, compressed yeast—a preserved ferment—does this read! How fortunate that its development demands abundant moisture, and that free aëration is as fatal to it as to the sulphuretted hydrogen exhalations on open salt marshes! Hence the absolute freedom of inland towns.

CHAPTER II.

PHYSICIANS in attendance on the sick, in epidemic times, have manifested almost as little acumen, in regard to the immediate source of pestilence, as success in treatment. This is not meant as a reflection on the medical profession, since, for a hundred years past, highly competent and judicious investigators have amassed the data and discussed opposite views with ardor, always supplying some facts with lengthened arguments, so as to enable the careful student, now, to take advantage of world-wide observations, accurate records of the course and symptoms of the disease, its field of progress, development, and limitation. The apparently interminable diversity of opinion is undoubtedly the basis of a unity which can not fail to strike us as growing and consolidating.

When I first investigated the lung-plague in cattle in the London dairies, the mortality was frightful and the proposed remedies were innumerable. Professional and non-professional observers were unanimous as to the origin of the disease in the London dairies, and those most active in attendance on the sick animals were apparently least competent to take a broad view of the nature and method of propagation, which Gerlach in 1836 found to be pure contagion, and which I afterward proved to be the case with this and other epizoötics, the concomitants of the free import of live animals into Great Britain, on the adoption of the otherwise glorious policy of free trade.

Sound pathological knowledge, guiding the statesman, from 1839 to 1846 and since, would have saved the British Isles from ten to twenty millions sterling per annum during nearly forty years. The British people have lost not less than £250,000,000 sterling, or, approximately, the appalling sum, in round numbers, of $1,200,000,000, by the direct and indirect effects of specific animal contagia.

Calculations have been made of the losses incurred by the United States from the yellow fever, and they are enormous. Conjectural as these figures may be, it would be well, some day, for a political economist to investigate this subject, especially should this be needed as an incentive in inaugurating so radical a policy, in relation to the shipping connected with the tropical Atlantic belt, as will guarantee the ports from disease invasion. The people of the United States are generous and humane, so that any statesman may count on national support who will take this subject in hand, and so coöperate with other governments as to accomplish the object in view thoroughly, not counting the cost.

Humanity, apart from any sordid considerations, demands the prompt and final extinction of yellow fever by all the means which science and a sound international policy may suggest. The helpless, dying seamen utter but faint cries, soon buried with their bodies in the deep. One crew is dispatched—another secured, to follow likewise; and a convalescent woman—a heroine —with a dying father, husband, or brother by her side, may have to navigate the ship, with none but almost images of death to sail and steer. We search for autochthonous causes of yellow fever on the banks of the Mississippi, as people were searching for them in Dr. Tully's days on the Connecticut. We so soon forget the signals of distress waving from pestilent vessels; which, helpless on the Gulf, have to be towed to quarantine. Thence germs, *if germs*—things, *if things* infected—find a readier entrance to the houses and stores of benighted Memphis than of the Crescent City.

A Plimsoll is wanted in America—one who can and will search out the conditions under which the shipping of the West Indies exists. The recklessness of ignorance must oft be added to the recklessness of intoxication, to induce even a brave young seaman to step into a charnel-house and cast his lot with the decomposing bodies of those still partially living. I have some idea of the horrible, fiendish purposes of man in trade, when time and danger have hardened him to unmanly deeds. Let us save the brave and inexperienced from any stealthy demoniacal assaults, and in doing this we shall reward the humane and careful—the model sea-captain and his cleanly crew—by depriving the diseases of the delta of the Mississippi and the waters beyond it of all their horrors—of their hold on human life and on the destinies of a glorious continent.

PAST HISTORY OF YELLOW FEVER.

My treatment of this subject can only be partial, and its object, so far as I can extend it, is to add convincing proof of the soundness of the view that yellow fever is a disease of ships in the tropical waters of the Atlantic. It is undeniable that it has a *permanent* home in the shipping, a *sub-permanent* home in seaports, and a *transient* home in river ports and inland towns, centers of commerce or the resorts of refugees.

Were I to undertake a purely medical history, I might divide the century since 1780 into the eras of fashionable medical opinions: the first, when the malady was deemed a visitation of God and springing anywhere; the days when Noah Webster wrote on Pestilence, when Pym, Devèze, Nathaniel Potter, and a host of others enriched the literature of yellow fever and developed the fight with the apostles of contagion; the second closing with the labors of Chervin, Louis Trousseau, and the eminent pathologists of that era; the third ending with Dr. La Roche's searching exposition of accumulated knowledge; and now, in the fourth quarter, the science of preventive medicine has made such strides that pills and potions are almost cast aside for the greater conquest of mind over matter—the control of conditions which may cure the city rather than the sick—save the ship and the seaman too. It is this proud end for which the foundation should practically be laid before the close of 1880, and I trust I am not deceived in believing that 1879 may be rendered memorable as the last year in which yellow fever had a firm hold on any important town in the United States.

Ancient History of Yellow Fever.

We know comparatively little of the ancient history of yellow fever. It was unknown before the discovery of the New World by Columbus. America was undoubtedly the seat of indigenous maladies—maladies actually springing from the soil, or *autochthonous ;* and we readily learn that they presented one common character, well understood by the modern pathologist. They were all such as are influenced in their periodicity by quinine. Yellowness, not necessarily jaundice, and hæmorrhages are common symptoms both of these endemics and of yellow fever ; black vomit—*vomito prieto* or *negro*—pertains occasionally to both.

The endemic character of yellow fever on land has met with the widest support, owing, in a great measure, to the comparative immunity enjoyed by the people of the West Indian Islands and Central American Atlantic shore during yellow fever outbreaks, and the virulence of the malady among strangers. That vaunted freedom is in a sense illusory. It is not unlike the asserted soundness of Texan herds, which destroy, by pasture contamination, northern cattle feeding with them. Texan bullocks thrive while 95 per cent. of animals grazing with them succumb to splenic fever with hæmorrhagic symptoms. On my visit to Texas I discovered that the vaunted immunity was a myth. Every animal bore marks of organic lesions which indicated a sub-acute and chronic state of sickness or previous attacks. Dr. Cornilliac * directs attention to the undoubted cases of yellow fever among creoles, and quotes Dr. Catel's observations in 1838, as also the statements of Professor Lufz, to the effect that in times of yellow fever the children of a town, and especially the new-born, pass through the malady, however mitigated it may appear and unobserved by the parents. The sallow and jaundiced look of the people of a malarial country is proof positive that their health is or has been impaired. Dr. Cornilliac asserts that the opinion respecting the identity of the nature of the fever which attacks children and indigenous adults during epidemics of yellow fever is accepted by all doctors in Martinique, where he has studied the disease.

* "Études sur la Fièvre Jaune à la Martinique," 2d ed., 1875.

Oviedo * reports that in 1494 a disease associated with jaundice appeared two months after the arrival of Columbus; and Gimaron alludes to the yellowness of the followers of Columbus: they were saffron-colored, or, as Herrera stated it, *azafranados*.

Diseases of the Red Men.

Dr. Hosack published letters from a trustworthy observer in the first volume of the "New York Medical and Physical Journal" (1827). This gentleman, Mr. John D. Hunter, says: "The diseases of the Indians are for the most part simple. . . . Those inhabiting wet, marshy land, especially such as reside on the shores of rivers which inundate their banks to a considerable extent, are somewhat subject, during the vernal and autumnal months, to bilious remitting and intermittent fevers; and those who have been reduced by the enervating corruptions of the frontier settlers have of late years been afflicted with typhus fever."

Rheumatism was the deadly disease which shortened life and rendered their brief span of existence a period of torment. They were excellent weather prophets, for their twinges and pains told of approaching atmospheric changes. They submitted to the most violent remedies. A hole was dug in the ground, a fire of dry sticks and brush was built over it, and on this water was poured, into which they entered as hot as it could be endured. Bushes and leaves were placed on them to retain the vapor, and, half suffocated, they remained there an hour or more. In Mexico to this day holes are dug in the ground by the Indians; stones are heated by fire and placed in the hole, and withdrawn at the right time; and the patient then enters and is packed in, keeping only the head out of water.

Reference has often been made to wide-spread disease among the Indians, both in the Northern States after the first arrival of Europeans and in the tropics. A very remarkable circumstance attending these outbreaks has been the ready propagation of the malady inland, its long continuance at all seasons and for several successive years, and the non-communication of the malady to the whites, notwithstanding that in all probability the disease, which may have been small-pox, was first introduced by a new-comer.

---

* " Historia General de las Indias."

Humboldt,* whose chapter on this question, extending over 38 quarto pages, is admirably lucid and terse, says: " Long after the arrival of Cortes an epidemic reigned periodically in New Spain, which the natives called *matlazahuatl*, and which authors have confounded with the *vomito* or yellow fever. This pest is probably the same as that which in the eleventh century forced the Kalteques to continue their migrations southward. It committed great ravages among the Mexicans in 1545, 1576, 1736, 1737, 1761, and 1762; but, as we have already indicated above, it was characterized by two features which were essentially different from those of the *vomito* of Vera Cruz: it attacked the natives and the copper-colored race almost solely, and it prevailed in the interior of the country at 1,200 or 1,300 feet above the sea-level. It is true that the Indians of the Mexican valley died by thousands in 1761, victims of the matlazahuatl, vomiting blood, which was ejected by the mouth and nose; but these gastric hæmorrhages (*hématémèses*) often occur in the tropics, accompanying the low bilious fevers. They have been likewise observed during the epidemic of 1759, which traversed all South America from Potosi and Oruro to Quito and Popayan, and which, according to Ulloa's description, was a *typhus* peculiar to the elevated regions of the Cordilleras."

Prescott distinctly avers that, when Cortes and his followers first entered the *tierra caliente*, in 1519, precisely the period when *vomito* is now known to have raged with greatest fury, no mention was made of an uncommon mortality. He says: " The fact doubtless corroborates the theory of those who postpone the appearance of the yellow fever till long after the occupation of the country by the whites."

I have striven to trace evidence of genuine yellow fever among the aborigines of America, and consulted Major Powell, whose knowledge of the Indians is surpassed by none; and there is absolutely no trustworthy evidence of ravages by any such disease. Decimation by small-pox has been common, and the pioneer settlers in Massachusetts, intent on converting the savage, asserted that the plague was evidence of God's wrath for the inhospitable treatment they had shown His chosen people, the

* " Voyage de Humboldt et Bonpland," p. 752, 3d part, vol. ii., Paris, 1811.

whites. When the Plymouth pilgrims arrived, the Indians feared a repetition of outbreaks, which were then on the decline.

During the reign of Charles V. of Spain, Velasquez, Governor of Cuba, determined on attacking Cortes on the Aztec coast, and he sent a force under the command of a Castilian hidalgo, Panfilo de Narvaez. The surprise and capture of the latter quickly followed, but Cortes took advantage of a formidable number of his foes by making them confederates, distributing gold and other precious commodities among the soldiers of Narvaez. Thus came the natives to experience far more injury from pestilence than war. A negro in Narvaez's suite brought with him the small-pox. " The disease spread rapidly in that quarter of the country, and great numbers of the Indian population soon fell victims to it." *

I note this well-authenticated case to show that contact with Europeans destroyed the Indians by other maladies than yellow fever—that disease of ships and seaport towns, and not of the wigwam.

Spanish and Portuguese Traders.

By a papal bull of Alexander VI. the discoveries in the Indies were made over to Spain and Portugal, but the difficulties attending the peopling and commercial development of the region led to licenses being given by the Spanish Government to private traders. As Prescott states,† the discoveries of Columbus resulted in a spirit which was evinced " in the alacrity with which private adventurers embarked in expeditions to the New World, under cover of the general license during the last two years of the [fifteenth] century. Their efforts, combined with those of Columbus, extended the range of discovery from its original limits, twenty-four degrees of north latitude, to probably more than fifteen south, comprehending some of the most important territories in the western hemisphere. Before the end of 1500 the principal groups of the West India Islands had been visited, and the whole extent of the southern coast, from the Bay of Honduras to St. Augustine. One adventurous mariner, indeed, named Lefe, penetrated several degrees south of this, to a point not reached by any other voyager for ten or

---

* " Conquest of Mexico," vol. ii., p. 272. Herrera, " Hist. General."

† " History of the Reign of Ferdinand and Isabella," p. 505.

twelve years after. A great part of the territory of Brazil was embraced in this extent, and two successive Castilian navigators landed and took formal possession of it for the crown of Castile, previous to its reported discovery by the Portuguese Cabral, although these claims to it were subsequently relinquished by the Spanish Government, conformably to the famous line of demarkation established by the treaty of Tordisellas."

It is asserted, and Cornilliac believes, that during all this period the yellow fever was the Indians' ally or auxiliary in their defense against the cruel cupidity of the Spaniards.

In Peru it has been noticed that the Indians are peculiarly liable to yellow fever, and do not enjoy the immunity of the negroes and Mongolians.*

The Spanish Galleons.

It is my belief that the Spanish galleon is the genuine ancestor of our plague-stricken ships. Dr. Joseph Jones, of New Orleans, † has rendered good service by directing special attention to this point. It is full of interest in tracing the first lines of the history of yellow fever. He says:

"At the time when Spain possessed by far the best and largest portion of the American Continent, extending from the north of California to the Straits of Magellan—a space of between six thousand and seven thousand miles—a system of commerce was established which appeared to be eminently favorable to the origin and spread of yellow fever. The Spanish galleons were, in fact, very large men-of-war, built in such a manner as to afford ample room for the stowage of merchandise, with which they were commonly so encumbered as to be rendered incapable of defense. The fleet of galleons consisted of eight such men-of-war, and generally convoyed from twelve to sixteen merchant-men. During times of peace the galleons sailed once a year regularly, though at no set time, but according to the pleasure of the King of Spain and the convenience of the merchants.

"They sailed from Cadiz to the Canaries, thence for the Antilles, and after reaching this longitude they bore away for Carthagena. As soon as they came in sight before the mouth of Rio de la Hacha, after having doubled Cape de la Vela, advice of their arrival was sent to all parts, that everything might be prepared for their reception. They remained a month in the har-

* See report by Dr. John M. Browne on the Oroya fever, "Medical Essays U. S. N.," Washington, 1872.

† "Proceedings of the Louisiana State Medical Association," New Orleans, 1879, p. 64.

6

bor of Carthagena, and landed there whatever was designed for *terra firma.* They then sailed to Puerto Velo, where, having staid during the fair, which lasted five or six weeks, they landed the merchandise intended for Peru, and received the treasures and commodities sent from thence. These galleons then sailed back to Carthagena, and remained there till their return to Spain, which usually happened within the space of two years. When orders for returning home arrived, they sailed first to the Havana, and having joined the flota, and what other ships were bound to Europe, they steered northward as far as Carolina, and then, taking the westerly winds, they shaped their course to the Azores, when, having watered and victualed afresh at Terceira, they thence continued their voyage to Cadiz.

"The Spanish flota consisted, like the galleons, of a certain number of men-of-war and merchant-ships; there were seldom more than three of the former and sixteen of the latter in this fleet. They sailed from the coast of Spain some time in the month of August, in order to obtain the winds that blow in November, for the more easy pursuing their voyage to Vera Cruz. They called at Puerto Rico on their way to refresh, passed in sight of Hispaniola, Jamaica, and Cuba, and, according to the winds and season, sailed either to the coast of Yucatan, or higher through the Gulf to Vera Cruz. The Spanish flotilla being intended to furnish not only Mexico, but the Philippine islands also, with the goods of Europe, was obliged to remain in Vera Cruz for a considerable time, and sometimes found it necessary to winter in that port. This fleet usually sailed from Vera Cruz in the month of May, but was sometimes detained as late as August; it then made for Havana, and returned to Spain in company with the galleons.

"The Spanish towns were generally built in low, unhealthy localities, surrounded by marshes and swamps, with narrow streets and high walls and fortifications, which not only compressed the towns within certain limits, and induced crowding and favored the accumulation of filth, but also prevented to a certain extent the free circulation of air."

Sir Walter Raleigh's Voyage.

When Columbus had shown the road, others were not slow to follow, and the British, French, Portuguese, and Dutch ships added to the facilities for the development of the tropical pestilence. We have to read the history of the seventeenth century to approach the conditions of common infection by yellow fever in tropical seas. One notable instance which bears strongly on my views of its development in mid-ocean is furnished by Sir Walter Raleigh's voyage in 1616. His fleet left England pure, healthy, and uncontaminated. It arrived off Lancerota on the 6th of September, and thence they steered to Gomera. After leaving this the weather was stormy, the ships were damaged,

perpetual rains and intolerable heat prevailed on their westward course. That they were steering close to the equator, toward Guiana, is proved by the high temperature encountered,
which bred disease on board, as usual in this region, destroyed
great numbers of men, and ultimately attacked the Admiral himself. He was seized with the most violent fever, and his recovery was despaired of. On the 12th of October, after having
been at sea over a month, they struck a dead calm, and toward
the end of the month their water was short. The sickness continued with great intensity among the crew, and at last on the
11th of November the fleet arrived off Guiana. That this, like
all other fleets afflicted with specific sickness in the tropical Atlantic, was stricken with yellow fever, I for one do not doubt.

William Piso, in his splendid "Natural History of the Brazils" (1648), writing "De Morbis Contagiosis," distinguishes the
endemics of the Brazils from the "maligna et contagios a febre"
attacking the Portuguese sailors and soldiers on the ocean. More
than a third of those proceeding from Europe to the Brazils died
in 1639. The long and tedious voyage, corruption of food, etc.,
induced the "morte repentina" characterized by blood extrava-
sations.

Without attempting a critical investigation of minor outbreaks of ocean pestilence, we may review the most interesting
of all the grand disasters assailing the fleet of one of the greatest of England's mariners.

ANSON'S VOYAGE ROUND THE WORLD.

Anson's "Voyage Round the World" is replete with interest,
and was published in London in 1749, the nominal editor being
Richard Walter, M. A., Chaplain to his Majesty's Ship Centurion. Benjamin Robbins, the author of "Mathematical Tracts,"
was in truth intrusted with working Mr. Walter's notes into
shape, and produced a work of great accuracy and repute. The
Spaniards adopted, as we have seen, a policy of exclusion in
their relations with the western islands and shores of the Atlantic, and such as would lead to conflicts and jealous aspirations.
In the summer of 1739 it was foreseen that a war with Spain
was inevitable, and Sir Charles Wager, deeply intent and active,
summoned Anson to the Admiralty. After some negotiations

as to the ships and force to be employed, the Duke of Newcas-
tle, Principal Secretary of State, delivered on the 28th of June,
1740, his Majesty's instructions to Mr. Anson. And here we
have a singular instance of that misdirection, so common in
starting important aggressive operations, which must so often
have failed but for the determined pluck and unflinching en-
durance of England's army and navy. George Anson was sent
to the South Seas with Chelsea Hospital invalids, and nearly
half of the five hundred veterans who were the fittest for duty
deserted at Portsmouth, and were replaced by two hundred and
ten marines, raw and undisciplined men, culled from the different regiments, who went on board on the 8th of August, ena-
bling the squadron to sail on the 10th of August, 1740.

The statement is made by the editor (and it is important in
such a record as this) that the Chelsea out-pensioners were op-
posed to and alarmed at the prospects of a voyage, "hurried
from their repose into a fatiguing employ, to which neither the
strength of their bodies nor the vigor of their minds was in any
ways proportioned, and where, without seeing the face of the enemy, or in the least promoting the success of the enterprise, they
would in all probability uselessly perish by lingering and painful
diseases." How true these anticipations were we shall soon see.

Ships composing the Squadron.

"The squadron allotted to this service consisted of five men-
of-war, a sloop-of-war, and two victualing ships. They were the
Centurion, of sixty guns, four hundred men, George Anson,
Esq., commander; the Gloucester, of fifty guns, three hundred
men, Richard Norris commander; the Severn, of fifty guns,
three hundred men, the Honorable Edward Legg, commander;
the Pearl, of forty guns, two hundred and fifty men, Matthew
Mitchel commander; the Wager, of twenty-eight guns, one hundred and sixty men, Dandy Kidd commander; and the Tryal
sloop, of eight guns, one hundred men, the Honorable John
Murray commander. The two victualers were pinks, the largest of about four hundred, and the other of about two hundred,
tons burden; these were to attend us till the provisions we had
taken on board were so far consumed as to make room for the
additional quantity they carried with them, which, when we

had taken into our ships, they were to be discharged." The 16 ships numbered 1,510 men on board.

After leaving St. Helen's Anson joined a convoy and traders, "making up eleven men-of-war and about one hundred and fifty sail of merchantmen, consisting of the Turkey, the Straits, and the American trades." On the 25th of October Anson's squadron and the two victualers cast anchor in Madeira, having long parted company with the rest. They took in water, wine, and provisions, made some changes in the officers, and on the 4th of November sailed, naming St. Jago, on the Cape de Verde Islands, as the first place of rendezvous in case of separation. If they did not meet the Centurion there, they were to make the best of their way to the island of St. Catherine's, on the coast of Brazil.

They first heard at Madeira of a cruising squadron, which they afterward ascertained was Don Joseph Pizarro's. A history of the Spanish squadron is given by the editor.* On the 21st of December Pizarro heard of the treachery of the Portuguese governor of St. Catherine's, of Mr. Anson's having arrived at that island on the 21st of December preceding, and of his preparing to put to sea again with the utmost expedition. Pizarro's fleet suffered seriously; and here the editor adds: "To all the misfortunes we had in common with each other, as shattered rigging, leaking ships, and the fatigues and despondency which necessarily attended these disasters, there was superadded, on board our squadron, the ravage of a most destructive and incurable disease."

Instead of touching at St. Jago, Anson ordered his fleet to the Island of St. Catherine's on the coast of Brazil.

* "This squadron (besides two ships intended for the West Indies, which did not part company till after they had left the Madeiras) was composed of the following men-of-war, commanded by Don Joseph Pizarro: The Asia, of sixty-six guns and seven hundred men, which was the Admiral's ship; the Guipuzcoa, of seventy-four guns and seven hundred men; the Hermiona, of fifty-four guns and five hundred men; the Esperanza, of fifty guns and four hundred and fifty men; the St. Estevan, of forty guns and three hundred and fifty men; and a patache of twenty guns. These ships, over and above their complement of sailors and marines, had on board an old Spanish regiment of foot, intended to reënforce the garrisons on the coast of the South Seas. When this fleet had cruised for seven days to the leeward of the Madeiras, they left that station in the beginning of November, and steered for the river Plate."

" On the 20th of November the captains of the squadron represented to the Commodore that their ships' companies were very sickly, and that it was their own opinion, as well as their surgeons', that it would tend to the preservation of the men to let in more air between decks; but that their ships were so deep they could not possibly open their lower ports. On this representation, the Commodore ordered six air-scuttles to be cut into each ship, in such places where they would least weaken it."

Anson's Fleet Surgeon's Opinion.

And here we have a singular instance of that prescience on the part of a skillful surgeon which might have saved many a fleet if attended to. He said: "And on this occasion I can not but observe how much it is the duty of all those who, either by office or authority, have any influence in the direction of our naval affairs, to attend to this important article, the preservation of the lives and health of our seamen. If it could be supposed that the motives of humanity were insufficient for this purpose, yet policy and a regard to the success of our arms, and the interest and honor of each particular commander, should naturally lead us to a careful and impartial examination of every probable method proposed for maintaining a ship's crew in health and vigor. But hath this always been done? Have the late invented plain and obvious methods of keeping our ships sweet and clean, by a constant supply of fresh air, been considered with that candor and temper which the great benefits promised hereby ought naturally to have inspired? On the contrary, have not these salutary schemes been often treated with neglect and contempt? And have not some of those who have been intrusted with experimenting their effects been guilty of the most indefensible partiality in the accounts they have given of these trials? Indeed, it must be confessed that many distinguished persons, both in the direction and command of our fleets, have exerted themselves on these occasions with a judicious and dispassionate examination, becoming the interesting nature of the inquiry; but the wonder is that any could be found irrational enough to act a contrary part in despite of the strongest dictates of prudence and humanity. I must, however, own that I do not believe this conduct to have arisen from motives so savage as the first reflection thereon does naturally suggest; but I rather impute it to an obstinate, and in some degree superstitious attachment to such practices as have been long established, and to a settled contempt and hatred of all kinds of innovations, especially such as are projected by landsmen and persons residing on shore. But let us return from this, I hope, not impertinent digression."

Outbreak of the Calenture.

They crossed the equator on the 28th of November, and on the 11th of December they were thirty-four leagues off Cape St. Thomas, right in the yellow-fever belt.

"We began to grow impatient for a sight of land, both for the recovery of our sick, and for the refreshment and security of those who, as yet, continued healthy. When we departed from St. Helen's, we were in so good a condition that we lost but two men on board the Centurion in our long passage to Madeira. But in this present run between Madeira and St. Catherine's we were remarkably sickly, so that many died, and great numbers were confined to their hammocks, both in our own ship and in the rest of the squadron, and several of these past all hopes of recovery. The disorders they in general labored under were such as are common to the hot climates, and what most ships, bound to the southward, experience in a greater or less degree. These are those kind of fevers which they usually call calentures; a disease which was not only terrible in its first instance, but even the remains of it often proved fatal to those who considered themselves as recovered from it. For it always left them in a very weak and helpless condition, and usually afflicted with fluxes or tenesmus. By our continuance at sea all these complaints were every day increasing, so that it was with great joy that we discovered the coast of Brazil on the 18th of December at seven in the morning."

What is a Calenture?

Dr. Ferguson, Medical Inspector of Hospitals, a most competent observer at the close of the last century, and well acquainted with the West Indian islands, explains this word calenture, which even Cornilliac has alluded to, and which was one of the earliest, possibly the earliest, name given to yellow fever by naval sur-

geons, who had never seen or heard of the disease before entering, for the first time, the deadly Atlantic belt. Dr. Ferguson
says "soldiers and others have been attacked and died of yellow
fever before they landed in the West Indies, or could be exposed to the influence of land miasmata in any shape." . . .
"From this it would appear that a calenture, the synocha of
Cullen (the pure offspring of heat, as pneumonia is of cold),
runs a course similar to the yellow fever—the same as the true
bulam."

The run from Madeira lay right across the yellow-fever
region, and the thirty-seven days in heavily laden ships, deep in
the water and crowded with men and provisions, developed
disease the nature of which is not doubtful. Special attention
must be paid to the acknowledged state of health at St. Helen's
and the disastrous fever which immediately followed.

Rest at St. Catherine's.

They anchored at 5 p. m. of the 18th off the N. W. part of
the Island of St. Catherine's. On the morning of the 20th the
squadron weighed and stood in; a pilot went on board, and
brought them to anchor in five fathoms and a half, in a commo-
dious bay, called Bon Port by the French. The next morning
they weighed anchor again, running between the castles of Santa
Cruz and St. Juan, and again anchored at one in the afternoon.
"In this position we moored at the Island of St. Catherine's on
Sunday, the 21st of December, the whole squadron being, as I
have already mentioned, sickly and in great want of refresh-
ments: both which inconveniences we hoped to have soon removed at this settlement, celebrated by former navigators for
its healthiness and the plenty of its provisions, and for the freedom, indulgence, and friendly assistance there given to the ships
of all European nations in amity with the crown of Portugal."

The bad reputation for unhealthiness of the equinoctial or
south seas has been referred to; and below Capricorn, where
the squadron was now anchored, the conditions for health were
better, but still bad for infected ships.

The narrative continues: "Our first care, after having
moored our ships, was to get our sick men on shore; prepara-
tory to which, each ship was ordered by the Commodore to erect

two tents: one of them for the reception of the diseased, and the other for the accommodation of the surgeon and his assistants. We sent about eighty sick from the Centurion, and the other ships, I believe, sent nearly as many in proportion to the number of their hands. As soon as we had performed this necessary duty we scraped our decks, and gave our ship a thorough cleansing; then smoked it between decks; and after all washed every part well with vinegar. These operations were extremely necessary for correcting the noisome stench on board, and destroying the vermin; for, from the number of our men and the heat of the climate, both these nuisances had increased upon us to a very loathsome degree, and, besides being most intolerably offensive, they were doubtless in some sort productive of the sickness we had labored under for a considerable time before our arrival at this island."

No one who has studied the history of yellow fever can doubt the nature of the evidence here before us; and it is especially valuable as proving the undoubted development of the plague at sea on perfectly healthy ships and with healthy people at the time of departure from St. Helen's.

The narrator speaks of the luxuriant soil of the island, fragrant woods, abundant fruits and vegetables, so that there is no want of pure apples, peaches, grapes, oranges, lemons, citrons, melons, apricots, and plantains. Onions, potatoes, cattle, and pheasants add to the abundance of the place; and the water on the island was excellent, preserving at sea as well as that of the Thames. These data are all of great interest in relation to the salubrity of the region, which was undoubtedly greater than in the yellow-fever equatorial zone. The editor says: "These are the advantages of this island of St. Catherine's; but there are many inconveniences attending it, partly from its climate, but more from its new regulations, and the late form of government established there. With regard to the climate, it must be remembered that the woods and hills which surround the harbor prevent a free circulation of the air. And the vigorous vegetation which constantly takes place there furnishes such a prodigious quantity of vapor, that all the night and a great part of the morning a thick fog covers the whole country, and continues until either the sun gathers strength to dissipate it or it is dispersed by a brisk sea-breeze. This renders the place close and humid, and probably occasioned the many fevers and fluxes we were there afflicted with."

The continued sickness, which was not at all to be wondered at, is ascribed to the unhealthiness of the place, the melancholy proof of which was that the Centurion buried twenty-eight men since her arrival, and the number of her sick increased from eighty to ninety-six, a proportion which would undoubtedly have been much greater had they anchored in the West Indies or been detained by storms nearer the equator.

The editor comments on Brazil, its gold and diamonds; on the Rio Grande and St. Catherine's, where during their whole stay they were refreshing their sick on shore, wooded and watered the squadron, and cleansed the ships; and Mr. Anson gave directions that the ships' companies should be supplied with fresh meat, and that they should be victualled with whole allowance of all kinds of provisions.

They left St. Catherine's on the 18th of January, and soon encountered a storm. The Pearl was nearly captured by the Spanish squadron. Captain Kidd died on the 31st of January, and Anson's squadron cast anchor on the 17th of February at St. Julian's. The Tryal had lost her mainmast, and had to be repaired. There is no mention whatever of sickness here, except that Captain Saunders, who was promoted to the command of the Tryal sloop, was dangerously ill of a fever on board the Centurion. Captain Kidd's death necessitated changes and promotions, and on the 27th of February the squadron weighed at seven in the morning and stood to sea.

On the 7th of March, in the morning, they opened the straits of Le Maire; and about 10 o'clock, the Pearl and the Tryal being ordered to keep ahead of the squadron, the straits were entered by the latter, with fair weather and a brisk gale. They all thought the greatest difficulty of their passage was at an end, and, animated by flattering delusions, they passed those memorable straits ignorant of the calamities then impending, and that the time drew near when the squadron would be separated, never to unite again. They experienced, on leaving the southern extremity of the straits, a violent storm. Doubling Cape Horn was discovered to amount really to an enterprise, and for the three succeeding months the distresses with which they struggled could not easily be paralleled by any former naval expedition. The tempests were dreadful and occurred at deceitful intervals. On the 18th of March cold was intense, and I need not detail the accidents, separations, and permanent losses sustained from this time till the month of May. I wish here to note that yellow fever was evidently at an end before they reached Cape Horn, and the winter weather had doubtless cleared the ships of every vestige of that disease. It is instructive therefore to see under what conditions, in these days when lime-juice and preserved vegetables were unknown as remedies for scurvy, this malady broke out.

Our author says : " Soon after our passing the straits of Le Maire, the scurvy began to make its appearance among us; and our long continuance at sea, the fatigue we underwent, and the various disappointments we met with, had occasioned its spreading to such a degree that at the latter end of April there were but few on board who were not in some degree afflicted with it, and in that month no less than forty-three died of it on board the Centurion. But though we thought that the distemper had then risen to an extraordinary height, and were willing to hope that as we advanced to the northward its malignity would abate, yet we found, on the contrary, that in the month of May we lost near double that number; and, as we did not get to land till the middle of June, the mortality went on increasing, and the disease extended itself so prodigiously that, after the loss of above two hundred men, we could not at last muster more than six foremast men in a watch capable of duty."

It is not my object to prolong this interesting chapter of the history of seafarers' diseases. On the 9th of June the Centurion arrived, at daybreak, at the island of Juan Fernandez, where the excellence of the climate, and the abundance of vegetable productions of all sorts, constituted the best possible remedies wherewith to recruit the squadron's remnant. Without relating the many brave deeds of the handful of men after leaving St. Juan, I may refer to the fact that the pink Anna had to be broken up as unseaworthy toward the end of August; and our author adds as follows :

"This transaction brought us down to the beginning of September, and our people by this time were so far recovered of the scurvy that there was little danger of burying any more at present; and therefore I shall now sum up the total of our loss since our departure from England, the better to convey some idea of our past suffering and of our present strength. We had buried on board the Centurion since our leaving St. Helen's 292, and had now remaining on board 214. This will doubtless appear a most extraordinary mortality, but yet on board the Gloucester it had been much greater; for out of a much smaller crew than ours they had lost the same number, and had only 82 remaining alive. It might be expected that on board the Tryal the slaughter would have been the most terrible, as her decks were almost constantly knee-deep in water; but it happened otherwise, for she escaped more favorably than the rest, since she only buried 42, and had now 39 remaining alive. The havoc of this disease had fallen still severer on the invalids and marines than on the sailors; for on board the Centurion, out of 50 invalids and 79 marines, there remained only 4 invalids, including officers, and 11 marines; and on board the Gloucester every invalid perished, and out of 48 marines only 2 escaped. From this account it appears that the three ships together departed from England with 961 men on board, of whom 626 were dead before this time; so that the whole of our remaining crews, which were now to be distributed among those ships, amounted to no more than 335 men and boys—a number greatly insufficient for manning the Centurion alone, and barely capable of navigating all three, with the utmost exertion of their strength and vigor. This prodigious reduction of our men was still the more terrifying as we were hitherto uncertain of the fate of Pizarro's squadron, and had reason to suppose that some part of it at least had got round into these seas. Indeed, we were satisfied from our own experience that they must have suffered greatly in their passage; but then, every port in the South seas was open to them, and the whole power of Chili and Peru would doubtless be united in refreshing and refitting them, and recruiting the numbers they had lost."

A most striking example of the vast difference there is in different seas, when an overcrowded ship is on a perilous voyage, may be adduced in the case of the Centurion, which fought the Spanish galleon, a rich prize, amounting to near $1,500,000, which she captured off Cape Espiritu Santo.

"The Commodore, when the action was ended, resolved to make the best of his way with his prize for the river of Canton, being in the mean time fully employed in securing his prisoners and in removing the treasure from on board the galleon to the Centurion. The last of these operations was too important to be postponed; for, as the navigation to Canton was through seas but little known, and where from the season of the year very tempestuous weather might be expected, it was of great consequence that the treasure should be sent on board the Centurion, which ship, by the presence of the commander-in-chief, the larger number of her hands, and her other advantages, was doubtless better provided against all the casualties of winds and seas than the galleon. And the securing the prisoners was a matter of still more consequence, as not only the possession of the treasure but the lives of the captors depended thereon. This was indeed an article which gave the Commodore much trouble and disquietude, for they were above double the number of his own people; and some of them, when they were brought on board the Centurion, and had observed how slenderly she was manned, and the large proportion which the striplings bore to the rest, could not help expressing themselves with great indignation to be thus beaten by a handful of boys. The method which was taken to hinder them from rising was by placing all but the officers and the wounded in the hold, where, to give them as much air as possible, two hatchways were left open; but then (to avoid any danger that might happen while the Centurion's people should be employed upon deck) there was a square partition of thick planks, made in the shape of a funnel, which inclosed each hatchway on the lower deck and reached to that directly over it on the upper deck. These funnels served to communicate the air to the hold better than could have been done without them, and at the same time added greatly to the security of the ship; for, they being seven or eight feet high, it would have been extremely difficult for the Spaniards to have clambered up; and, still to augment that difficulty, four swivel-guns, loaded with musket-bullets, were planted at the mouth of each funnel, and a sentinel with lighted match was posted there ready to fire into the hold among them in case of any disturbance. Their officers, who amounted to seventeen or eighteen, were all lodged in the first lieutenant's cabin under a guard of six men; and the general, as he was wounded, lay in the Commodore's cabin, with a sentinel always with him. Every prisoner, too, was sufficiently apprised that any violence or disturbance would be punished with instant death. And, that the Centurion's people might be at all times prepared, if, notwithstanding these regulations, any tumult should arise, the small arms were constantly kept loaded in a proper place, while all the men went armed with cutlasses and pistols; and no officer ever pulled off his clothes when he slept, or when he lay down omitted to have his arms always ready by him.

"These measures were obviously necessary, considering the hazards to which the Commodore and his people would have been exposed had they been less careful. Indeed, the sufferings of the poor prisoners, though impossible to be alleviated, were much to be commiserated; for, though the weather was extremely hot, the stench of the hold loathsome beyond all conception, and their allowance of water just sufficient to keep them alive, it not being practicable to spare them more than at the rate of a pint a day for each, the crew themselves having only an allowance of a pint and a half—all this considered, it is wonderful that not a man of them died during their long confinement, except three of the wounded, who expired

the same night they were taken, though it must be confessed that the
greatest part of them were strangely metamorphosed by the heat of the
hold; for when they were first brought on board they were sightly, robust
fellows; but, when, after above a month's imprisonment, they were discharged in the river of Canton, they were reduced to mere skeletons, and
their air and looks corresponded much more to the conception formed of
ghosts and specters than to the figure and appearance of real men."

Here have we a splendid series of examples, indicating, first,
the development of mortal yellow fever when the squadron
sailed westward from Madeira, crossing the line and sailing
slowly southward to St. Catherine's Island; secondly, of scurvy
by improper diet and privations after rounding Cape Horn;
and thirdly, the complete immunity with which a ship which
had been badly infected with yellow fever (subjected, it is
true, to severe frost the first winter after the outbreak) conveyed its crew and a host of prisoners across the Pacific Ocean.
They suffered all the horrors of a short supply of water, and
foul stenches from enforced imprisonment of crowds in the
hold, and yet reached Canton without the loss of a single life
by disease, such as would undoubtedly have decimated them in
the American archipelago.

Captain Cook's Voyage.

The history of Cook's celebrated voyage from 1772 to 1775
affords an instructive and never-to-be-forgotten contrast to Anson's disasters from pestilence. Commanding H. M. ship Reso-
lution, he had with him a company of 118 men, performing a
voyage of three years and eighteen days throughout all the climates from 52° north to 71° south, with the loss of only one
man by sickness. "How great and agreeable, then, must our
surprise be," said Sir John Pringle when presenting Sir God-
frey Copley's gold medal, at a meeting of the Royal Society,
to Captain Cook, "after perusing the histories of long naviga-
tions in former days, when so many perished by marine diseases,
to find the air of the sea acquitted from all malignity; and, in
fine, that a voyage round the world may be undertaken with
less danger perhaps to health than a common tour in Europe."

To this day has the illustrious Cook's lesson proved almost a
dead letter. He had the advantage of some, but still slight,

knowledge of antiscorbutics. He established a system of three watches for the crew; the sailors generally had dry clothes to put on when they happened to get wet, and every precaution was taken not to expose them unduly to wet. Cook did not believe in seasoning his men by rough and debilitating means. They were compelled to keep clean not only in their persons, but with their hammocks, bedding, and clothes. *The ship was kept clean and dry between decks.* Fires were used for airing, or the ship was smoked with gunpowder, vinegar, and water. "I had also," says Captain Cook, "a fire made in an iron pot at the bottom of the well, which was of great use in purifying the air in the lower parts of the ship. To this and to cleanliness as well in the ship as among the people, too great attention can not be paid; the least neglect occasions a putrid and disagreeable smell below, which nothing but fires will remove." Facts enabled Cook exultingly to declare "that our having discovered the possibility of preserving health, among a numerous ship's company, for such a length of time, in such varieties of climate, and amid such continual hardships and fatigues, will make this voyage remarkable in the opinion of every benevolent person, when the disputes about a southern continent shall have ceased to engage the attention and to divide the judgment of philosophers."

The eloquent Pringle, addressing the Royal Society as he handed Captain Cook the medal, said: "If Rome decreed the civic crown to him who saved the life of a single citizen, what wreaths are due to that man who, having himself saved many, perpetuates in your 'Transactions' the means by which Britain may now, on the most distant voyages, save numbers of her intrepid sons—her mariners who, braving every danger, have so liberally contributed to the fame, the opulence, and to the maritime empire of their country."

I have purposely enforced this lesson by Sir John Pringle's words, since history has to instruct us, not only concerning the ravages of plagues, but on the conditions under which these are prevented.

Bryan Edwards on West Indian Fever.

Numerous are Edwards's allusions to this subject, and in one he says : *

"Four years have elapsed since I closed the details of the military operations of the British army in St. Domingo, and I grieve to say that what
was then prophetic apprehension is now become historical fact. This once
opulent and beautiful colony, the boast of France, and the glory of the new
hemisphere, is expunged from the chart of the civilized world! The pros-
pect of such lamentable ruin might give occasion for many observations
and reflections; and I shall present to my readers, in the following very
imperfect sketch (for such it is in every sense), a few that occur to me;
more than this I dare not attempt. Were it in my power (as in truth it is
not) to continue in a regular series the history of those sad events which have
led to this miserable catastrophe, I should indeed decline a task which would
be equally disgusting to my readers and painful to myself. In a climate
where every gale was fraught with poison, and in a contest with uncounted
hosts of barbarians, what could the best efforts of our gallant countrymen
effect? Their enemies indeed fled before them, but the arrows of pestilence
pursued and arrested the victors in their career of conquest! Scenes like
these, while they afford but small cause of gratulation to the actors themselves, furnish no topics to animate the page of the historian, who would
have little else to display but a repetition of the same disasters—delusive
promises, unrealized hopes, unavailing exertions, producing a complication
of miseries, disease, distraction, contagion, and death!

"At the same time (although I know not that the reader will derive any
great degree of consolation from the circumstance) it is incumbent on me
to observe that, during the disastrous period of which I treat, I have not
heard that any misconduct or neglect was ever fairly imputed to those persons who had the direction of the enterprise, either in the public depart-
ments of Great Britain, or in the scene of action itself. The names of
Williamson, Forbes, Simcoe, Whyte, and Maitland carry with them a demonstration that neither courage, nor energy, nor military talents were at any
time wanting in the principal department. Reënforcements of troops, too,
were sent by the British Government with a more liberal hand than in former years. Toward the latter end of April, 1795, the 81st and 96th regiments (consisting together of 1,700 men) arrived from Ireland; the 82d,
from Gibraltar, landed 980 men in August; and in April, 1796, the 66th
and 69th regiments, consisting of 1,000 men each, with 150 artillery, arrived from the same place, under the command of General Bowyer; so that
the whole number of effective men which had landed in St. Domingo down
to this period (including some small detachments sent up at different times
from Jamaica) amounted to 9,800. In June following, four regiments of

* "History of St. Domingo," pp. 383–386, London, 1801.

infantry, and a part of two others, arrived from Cork, under the command of General Whyte. These were soon afterward followed by seven regiments of British together with three regiments of foreign cavalry; besides two companies of British, and a detachment of Dutch artillery; making in all a further reënforcement of about 7,900.

"But what avail the best concerted schemes of human policy against the dispensations of Divine Providence? A great part of these gallant troops, most of them in the bloom of youth, were conveyed, with little intermission, from the ships to the hospital—from the hospital to the grave! Of the 82d regiment, no less than 630 became victims to the climate within the short space of ten weeks after their landing. In one of its companies no more than three rank and file were fit for duty. Hompesch's regiment of hussars was reduced in little more than two months from 1,000 to 300, *and the 96th regiment perished to a man!* By the 30th of September, 1796, the registers of mortality displayed a mournful diminution of no less than 7,530 of the British forces only; and toward the latter end of 1797, out of the whole number of troops, British and foreign, which had landed and were detained in this devoted country, during that and the two preceding years (certainly not far short of 15,000 men), I am assured that not more than 3,000 * were left alive and in a condition for service.

"'In these adventures,' observes Mr. Burke, 'it is not an enemy we have to vanquish, but a cemetery to acquire. In carrying on war in the West Indies, the hostile sword is merciful; the country itself is the dreadful enemy; there the European conqueror finds a cruel defeat in the very fruits of his success. Every advantage is but a new demand for recruits to the West Indian grave.' Let us also hear on this subject the Poet of the 'Seasons':

> "'Then wasteful forth
> Walks the dire power of pestilent disease,
> Sick nature blasting, and to heartless woe
> And feeble desolation casting down
> The towering hopes and all the pride of man!
> Such as of late at Carthagena quenched
> The British fire.
>   Gallant Vernon saw
> The miserable scenes,
> Heard nightly plunged amid the sullen waves
> The frequent corse!'
>   (THOMSON.)

"This miserable scene, however," says Edwards, "has been frequently repeated since the siege of Carthagena, where the disease was imported from the West Indian Ocean. It was exhibited at the Havana in 1762; at the River St. Juan, and lately in the Windward Islands; but nowhere, I believe, with greater force and effect than in St. Domingo."

* "The loss of seamen in the ships employed on the coast is not included. It may be stated very moderately at 5,000 men."

7

Yellow Fever in British and French Guiana.

There are three centers of epidemic outbreaks of yellow fever which, without special reference to chronological order, tend to throw light on the conditions existing in the most insalubrious ports, and which have, to some extent, misled the most skillful observers.

Dr. Daniel Blair published a work on yellow fever as it invaded British Guiana. This is a low, flat, very marshy country, lying between 6° and 8° north latitude, having a uniform high temperature and humidity, and its climate has obtained a very ill repute from the ravages of this epidemic.

Yellow fever had been in Georgetown in 1819, but its characteristics had been forgotten; it had passed from memory, and the harbor had acquired a high character for healthiness. Nevertheless, the mercantile part of the town was low, overcrowded, and in far too close proximity and too nearly on the level with the shipping. Water Street and Robb's Town are embraced in this region. The "mud-lots" of Water Street, formed by the embankment that prevents the overflow of the river-tide, have attached to each a "stelling," or landing-wharf, extending beyond the buildings into the shelving clay-bed of the river. Seven public stellings, at several intervals, keep up a free communication between the city and the shipping. Across Water Street six sluices discharge the drainage and sewerage past the stellings into the river. The bulky materials of any kind floating among the piles are necessarily entangled and detained below the stellings. Over some of the stellings, where the water is quiescent, the most offensive smells arise, *and the white paint of the wooden houses is speedily reduced to metallic lead.*

It was on the mud-lots of Georgetown and their immediate neighborhood, close to those stellings where ships for the West Indies lie alongside, that a number of cases, soon recognized as mortal yellow fever, were recognized by the physicians of the city, including Dr. Blair. The importation by the shipping was not traced nor apprehended. But, for all that, Blair did not overlook, in the low room where he witnessed his earliest case, that it was a "filthy hole," with a sickening smell of rotten salt fish, tobacco, damp, and dirt. This original neighborhood, most accessible to the shipping, maintained its virulence during the epidemic, though this was most prevalent at the very mouth of the river, the moorings most sought after by shipmasters on account of the advantages of the free, open breeze.

Till 1842 Dr. Blair had charge of the plantation hospitals, and he noticed that, in proportion as the immigrants approached town or coastward in their locations, the yellow fever predominated, and as they receded the intermittent fever asserted supremacy. Whatever might be the cause inducing the epidemic, Dr. Blair had sufficient evidence to show that inquiries in this matter must be directed to the *shore*. " Some new element," he says, " is required in the generation of yellow fever besides what is to be found usually within our embankments, and it is in all probability dependent on a sea change."

The disease was recognized as not contagious. Dr. Blair declared that opinion in Demerara as " totally obsolete." In 1852 he witnessed another epidemic, and speaks of cases in his private practice as occurring in the neighborhood of Robb's stellings.

I may incidentally mention that a careful account has been given by a French naval surgeon, Dr. Lucien Richepin,* of the introduction of yellow fever in French Guiana in 1872; and his data assured him, and *"for his part he was profoundly convinced,"* that the yellow fever which broke out in 1872 in Guiana was developed in the hold of the Topaze. Men on board, soldiers and sailors, were there infected, *"and the seeds of the disease there fructified as on a propitious soil."*

He cites a case from the second volume of the " Archives of Naval Hygiene," for 1864, which establishes propagation by clothing. Six men out of eleven, who lived with the paymaster, were seized with yellow fever, and in this paymaster's house were deposited the clothes of the soldiers who had died in the hospital.

Another instance is most instructive, showing the influence of atmospheric communication or isolation. At Tampico a company of soldiers was lodged in barracks contiguous to the hospital, but separated by a high wall. For two months there

* " Thèse pour le Doctorat en Médecine," Paris, 1875.

was complete immunity. Windows were then made in this wall, and the soldiers, so far spared, were attacked in a few days, and decimated by the plague.

Yellow Fever on the African Coast.

If we cross the Atlantic in the same degrees of latitude precisely as Guiana, we find another clinching group of facts which support most strikingly the naval origin of yellow fever. We do not lack data. A high authority, Mr. R. Clarke, late surgeon to the natives on the west coast of Africa, contributed a valuable paper to the Transactions of the Epidemiological Society in 1863. I need not detail the conditions which render this region so deadly to Europeans. He tells us, "the epidemic or yellow fever, which at uncertain intervals scourges the colonies of Sierra Leone and the Gambia, *is wholly unknown on this coast.*" Speaking of Cape Coast Castle, which was built by the Portuguese in 1612, and has been held by the British since 1672, he says: "The removal of the jungle and the tillage of the waste lands in the neighborhood of Cape Coast would free it from a perpetual source of febrile and dysenteric diseases, which at all times more or less afflict its inhabitants." The sanitary condition of the town of Cape Coast is deplorable. No public *cloacinœ;* foul stenches everywhere; by-paths and beach offensive from accumulations of human ordure or filth; turkey-buzzards, many curs, and ill-conditioned hogs, devour the excrementitious matters which are left to rot upon the street; a ravine—a vast surface-sewer—is strewed over with animal and vegetable refuse in every stage of decay; at the seaside where it terminates, the surface-water is dammed up by the sand thrown up by the surf on shore, and is there collected into fetid pools. The houses in Cape Coast are so closely connected that a free current of air between them is much impeded; the population is condensed, for a number of families occupy each hut or house. The mass of the inhabitants *bury their dead in the basement floor of their houses*, a practice *not* confined to the pagan part of the population, but also practiced by many respectable and wealthy families. Precious relics and gold-dust are buried with the dead, and in times of trouble the tombs are opened, and the gold so deposited is applied to meet pressing claims. Mr. Clarke incidentally states that King David had a large treasure placed in his tomb by his son Solomon.

During a service of nearly twenty-three years' duration at Sierra Leone and the Gold Coast the diseases afflicting Europeans were seen by Mr. Clarke among the natives—fevers general slight; intermittent and typhoid prevalent. Yellow fever never attacked the natives in 1837, 1838, 1839, 1847 or 1859. It is confounded with small-pox in records of deaths, and this has led to confusion.

Elephantiasis and lepra in all its hideous forms prevail:

Sierra Leone is well known as possessing a most insalubrious climate. It is a *terra inhospita* for all Europeans who venture to reside in it, but it has suffered from yellow fever only in the way that this disease has attacked at intervals the several West Indian islands.

From the Army Medical Department Reports we learn of a severe epidemic in 1815, but the first authentic notice of yellow fever in the colony is due to Copeland, who mentions that several cases occurred among *the seamen*. The garrison remained exempt. In 1822 the first epidemic at Freetown occurred, and was ascribed to the captain of a merchant-vessel who took ill after having piloted H. M. S. Ranger into harbor. The timber vessels were the most affected.

Space forbids my following the recorded outbreaks from British official documents or the really classical works of Dr. Bérenger Féraud on yellow fever in Senegal. The history is a repetition of the history of the disease in the West Indian islands, but it illustrates most forcibly the manifestations of the malady primarily in the shipping and most frequently in the timber-laden mercantile craft; thence the disease enters the islands and seaports, having reached Cape Coast Castle and Elmira during the Ashantee campaign in 1873 (as any one might have foreseen it probably would), uniformly sparing the negroes, and never traveling, much less originating, beyond the confines of the pestilent cities.

Yellow Fever in Charleston.

I am indebted to Dr. J. M. Toner, of Washington, for his valuable essay on "the Natural History and Distribution of Yel-

low Fever in the United States from 1868 to 1874." In this he
points out very distinctly the influence of elevation on the disease and its preference for the low lands of the Gulf States and
the Atlantic coast—density of population or crowding in cities
favoring the spread of the disease.

Charleston, South Carolina, is only ten feet above the level
of the sea, and has been more frequently affected, since 1699,
than any other city in the United States, excepting New Orleans.
It has had fifty-three epidemics in 220 years, whereas New
Orleans has had 74 since 1769. Charleston has been singularly
exposed as a low-lying port readily accessible to vessels from
the yellow-fever belt and bathed by ocean water, so favorable
to the disease, whereas New Orleans is on an inland river. The
much more extensive trade of the latter and its position account
for its frequent infection since its first epidemic 110 years since.

If there be any city in the United States where the local
origin of yellow fever might by some be deemed possible, it is
Charleston ; and probably the same conditions exist nowhere
else. It is easy, however, to see that these features account
more readily for the production of the disease in ships in the
harbor than in the city to which it has so often been communicated from the shipping.

I have had the privilege of perusing a very interesting report of Dr. A. N. Bell, whose investigation of the tidal drains
of Charleston is very suggestive in relation to the influences
favoring the inroads of the disease.

The city is built on quicksand. The old drains had been
laid down without any definite plans on different levels, and
ending in *cul-de-sacs*. Dr. William I. Wragg was the author of
the tidal drains, built at a dead level 30 inches above mean low
water, and the top of the arch a foot above ordinary high water.
Water-tight doors were provided at each end, so that at the
highest tide those at the upper end would be opened for a
couple of hours or more to admit of filling the drains ; then
these doors were closed, and at the next low tide the lower gates
would be opened to empty the drains. The wash was always
in one direction, and, so long as a special drain attendant was
employed, the stagnation of sea-water and organic refuse was

never allowed to undergo the deleterious changes which would obviously occur with time and high temperature.

The drain-keeper was dismissed, and the police permitted to undertake and neglect his duties on a change of administration, after the tidal drains were in fair working order. The result has been the filling up of these dead-level passages, which, contrary to the original intent, have been made the receptacles for the sewage of houses and public institutions.

Dr. Bell, well acquainted with the special evils attendant on stagnant sea-water charged with excreta, shows that the tidal drains were solely adapted to surface and soil water. They are certainly the most dangerous devices of man in such a latitude and such a port as Charleston. They are now almost completely obstructed, and a source of accumulating danger. The letting in of the tide at high water and opening the gates at low water now flushes nothing. It merely saturates the contained mass, trickles out equally at both ends, and saturates the contiguous soil by leakage. The filth in the slips and on the river front is exceedingly offensive, and a coating of lead paint is blackened in a night. Sulphuretted hydrogen is there evolved where the charged sea-water is imperfectly aërated, and there seems to be fair ground for the belief that these conditions, which in a ship at sea or in a foul harbor in the West Indies will engender yellow fever, may have been sufficient to engender a fatal ship fever and possibly yellow fever in Charleston harbor.

This is related in the "Official Report" of Dr. Robert Lebby for 1876. In that year, "the first case of fever which approximated yellow fever occurred on board a schooner laden with coal *from Philadelphia* at Johnson's wharves, on Cooper River, northeastern section of the city. The schooner had been at this wharf for several days, and the mate sickened August 30th and died September 3d. A second case of a mild type was a clerk of the W. P. Hall on Brown's wharf. Again, on board the barque Sylph, which discharged her ballast of rock and sand, and, loaded with naval stores, hauled out into the stream off East Battery, the first mate sickened that night, then the second mate, and afterward two boys. . . . . The two mates died, and the boys recovered. Several vessels at Marshall's wharf, and other wharves near the outlets of large drains, had their crews attacked with this fever. From all the information I have been able to collect, I am of opinion that the fever was of local origin."

Dr. Bell continues: "In one of these slips, and on one of the hottest days of my inspection in August, I observed dredging going on at low water, with the view of getting rid of the liquid filth dammed up by the solid deposit at the outer margin. The loaded flat was pulled off to the edge of the channel only, and the load there dumped, in position to be swept back by the tide to the place from which it was taken, or into neighboring slips."

"Many of the houses of the poor people are devoid of cisterns, and the only drinking water is that which is supplied by shallow wells, virtual drain-holes of the foul premises. On inquiry among the families, I learned that their dead children were more numerous than their living ones."

Sullivan's Island and Mount Pleasant—summer resorts composed of sand-dunes; a very small portion turfed and few shade trees—the sandy surface the "catch-all" of all the filth—when the air is still evolve the characteristic odor of a dung heap. Mount Pleasant is under better cultivation than Sullivan's Island. Both commonly subject to yellow fever whenever it prevails in Charleston.

This apparent exception to my view of the impossibility of engendering the disease on land may turn out to be one of the clear indications of the soundness of the doctrine I am inculcating. The coal-ship in the reeking harbor under the influence of protracted heat and humidity could engender yellow fever in August as readily in Charleston as in the port of Kingston, Jamaica, or Havana, where such events have no doubt been exceedingly common. A ship entering Charleston with putrid bilge-water, and contaminated by infiltration of the putrescent mixtures in the Charleston slips, would constitute an admirable pest-trap, and the means of propagating infection when once the malignant ferment had gained supremacy.

European Outbreaks of Yellow Fever.

Forgetting chronology, here again I desire to show that the ports affected on the ocean shores, or the towns on the Spanish and French rivers, have derived all their yellow fever from the tropical Atlantic.

The earlier invasions of this century secured to science many excellent contributions, especially from French and British physicians. The disease afflicted many ports, such as Cadiz, Carthagena, Gibraltar, Leghorn, and Malta, and harassed the French and British fleets and troops in Egypt.

To Baron Larrey we owe one of the most singular—and in a pathogenic sense—one of the most instructive records of this plague. It seems to have spread among the wounded like erysipelas or septicæmia, and we must not forget that it is a putrefactive disease. Larrey* relates that the mortal accidents which afflicted the wounded, after the battle of Heliopolis and the siege of Cairo in 1800, led the French soldiers to fear that the enemy's balls had been poisoned. It was easy to dissuade them of this, but not so easy to arrest the effects of the malady. It appeared on the 5th of April and lasted to the end of May, attacking only the formidably wounded. It was a genuine yellow fever. Two hundred and sixty wounded of all kinds died with this complication, out of about six hundred from the siege of Cairo to the taking of Boulâg.

Dr. Savaresz, who saw the yellow fever in the Antilles, recognized the disease attacking the wounded as *typhus icterodes* of the English and French nosologists—the *vomito prieto* of the Spaniards.

The English who followed the French in Syria and Egypt suffered of yellow fever in their ships.

Recent Outbreaks in France and England.

The activity of commerce between England, France, and the West Indies has, at times, been attended by the accidental arrival of cases of yellow fever, further north on the European than on the American Atlantic shores. Dr. F. Mélier has, by his work on "Yellow Fever at St. Nazaire," rendered memorable an outbreak at this port, on the Loire. He has pointed out, in the first place, that the majority of the European outbreaks of yellow fever have been due to sugar ships. In the special invasion he reports, the focus of the malady was in the ship itself, and the hold was

* "Mémoires de Chirurgie Militaire et Campagnes." Paris, 1812.

its seat. So long as this hold remained closed, a very limited number of accidents occurred in crossing the Atlantic, but it was on opening the scuttles and hatchways that the disease spread in the port, in the ships and among people engaged on the infected vessel. This might be compared to a lethal weapon which exploded, killing all within its reach. The delinquent ship, the Anne Marie, had left Nantes for Havana, which it reached in ballast on the 12th of May, 1861. On the 13th of June she left with all hands healthy and a cargo of sugar; she passed through Florida Straits, and was becalmed twelve days. The heat was intense, and abundant rains and tempests contributed to render the passage tedious. It was *seventeen* days after leaving Havana before a man became sick. A young sailor, nineteen years old, was seized on the 1st of July. Later the same day another was attacked, and the two died on the 5th. A third case occurred on the 2d, and so forth till eight sailors were down and then the captain took it. Nine cases altogether occurred among sixteen persons, and two died. Twenty days after the last death, on the 25th of July, the vessel reached St. Nazaire, and from the length of time without a case no quarantine was enforced. I shall not detail the cases which occurred in two ships which were moored near the Anne Marie, or on land among persons who frequented or came near the vessel. A limited outbreak was the result in the northern parts of France.

Dr. George Buchanan was called upon officially to report on an outbreak of yellow fever at Swansea, in South Wales. Latitude 51° 37′ north, and longitude 3° 55′ west. Population in 1861, 42,581. The outbreak was unprecedented, and Mr. John Simon has epitomized as follows: "The broad facts of the case may be told in very few words under the following two heads: First, the Hecla left Cuba on the 26th of July with cases of yellow fever on board, had successive new cases till toward the end of August, entered Swansea harbor on the 9th of September, with one of her seamen dying and two others but convalescent from the fever, and was immediately moored alongside a wharf, where she landed her sick, discharged (though not uninterruptedly) her cargo, and remained stationary till the 28th; when remonstrances, which at last had become irresistible, led to her being removed from within the dock. Secondly, from September 15th, six days after her arrival, to October 4th, six days after her removal, Swansea witnessed the entirely new phenomenon of yellow fever attacking in succession some twenty inhabitants of the town, besides others who suffered less definitely, or more mildly; and this not indiscriminately over the whole large area of Swansea, but only in definite local relations to the ship; while at Llanelly there also fell sick in the same way three of the crew of a small vessel which had been lying alongside the Hecla at Swansea."

Still more recent but isolated cases have occurred, but I shall notice only one which enforces the lesson that a disease which can originate at sea need not be referred to any port the ship has touched.

Clean Bills of Health.

In the "Lancet" for May 12, 1877, a case is reported of a sick German seaman landed from the brig Däring, which had arrived at Salcombe from St. Lucia. The captain had a clean bill of health from the latter port, which may have been quite correct if the disease developed at sea, and Messrs. Cornish and Webb, medical officers of the union, recognized all the symptoms of yellow fever in their patient, who died after removal to the Hospital of the Kingsbridge Workhouse. When the nature of the disease was discovered, all other patients were removed and the building submitted to a process of disinfection.

With this I shall rest satisfied, and merely indicate by a rapid survey of the West Indian Islands how singularly their experience tallies with the observations made at New Orleans, Charleston, or other ports where we now know the disease never does and never can originate spontaneously within the city precincts.

SPECIAL DATA RELATING TO THE WEST INDIAN ISLANDS.

The island of Martinique, in 14° north latitude, has been afflicted most frequently with yellow fever. Its history in a sense indicates why it should, for it has been the center of very active operations, before and since the creole was known. If yellow fever springs from the soil anywhere, it is in the island of Martinique; but every autochthonous disease on the globe, that has yet been described with any precision, has been traced to a limited area marked by special conditions, as in the case of ague, cholera, and the more recent Oroya fever. The disease always continues somewhere within the belt of its origin, at its appropriate season and under well-known conditions. It is true, taking the widespread anthrax of animals, or malignant pustule in man, that it shifts its ground; and in the hot, dry season, when it prevails on clayey soils, it does not appear on less retentive lands; and *vice versa*, in warm, wet years, it spares the heavy clays.

All familiar with similar interesting maladies in men and animals know that the circumstances under which the indigenous diseases are made manifest, or attain exceptional virulence, are known, and constitute the basis of well-directed precautionary measures. But I challenge any one to glean, from the many volumes which a century has produced relating to yellow fever, anything like so definite a statement as can be made off-hand relating to endemic and enzoötic affections scattered over the whole world.

Let any one consult the annexed table, which I have prepared after Cornilliac. There are two outbreaks in the seventeenth century—one short, in 1669, and the other long, of 24 years, with an intervening period of 12 years, during which the island was absolutely free. The years when it was only partially free are those figured in italics, and the number is set out in the right-hand columns. During the eighteenth century we have four epidemic periods, of 13, 6, 12, and 17 years respectively; but the periods of immunity are actually longer, viz., 15, 14, 6, and 19 years respectively. With the more active commerce of the nineteenth century, we count from 1810 to 1873 four outbreaks, the duration of which was 12, 7, 8, and 5 years, and with this aggravating sign, that they were years without any seasonal intermission — no respite, but continuous mortality. The periods of immunity are also four, viz., two of 8, a third of 6, and a fourth of 10 years, when the island enjoyed perfect health.

Admitting, for argument's sake, that this chart may not be absolutely accurate, it is certain that other West Indian islands of bad repute, and indeed all of them, manifest similar and more startling contrasts, which prove, with absolute and incon-

PERIODS OF YELLOW-FEVER EPIDEMICS, AND OF COMPLETE IMMUNITY, OBSERVED AT MARTINIQUE.

PERIODS.	EPIDEMIC PERIODS AND PERIODS OF IMMUNITY. The figures in *italics* represent the year of intermission when the plague disappeared for several months in the year.	NUMBER OF EPIDEMIC YEARS.			NUMBER OF YEARS OF IMMUNITY.
		Without Respite.	With Seasonal Abatement.	Total.	
17th CENTURY, { 1st Epidemic Period.. Immunity.............	1669........ 1670 to 1681 inclusive.	1		1	12
2d Epidemic Period..	{ *1682, 1683, 1684, 1686, 1687, 1688, 1689, 1690, 1694, 1692,* 1693, 1691, 1695, 1696, 1697, 1698, 1699, 1700, 1701, 1702, 1703, 1704, 1705, 1706 ...... }	12	12	24	
{ Period of Immunity..	{ *1707, 1708, 1709, 1710, 1711, 1712, 1713, 1714, 1715, 1716,* 1718, 1719, 1720, 1721, 1722....... }				15
3d Epidemic Period..	{ *1723, 1724, 1725, 1726, 1727, 1728, 1729, 1730, 1731, 1732,* 1733, 1734, 1735....... }	3	10	13	
Period of Immunity..	1736 to 1749 inclusive......				14
4th Epidemic Period..	*1750, 1751, 1752, 1753, 1754, 1755.*	1	5	6	
Period of Immunity...	1756 to 1761 inclusive......				6
18th CENTURY, 5th Epidemic Period..	{ *1762, 1763, 1764, 1765, 1766, 1767, 1768, 1769, 1770, 1771,* 1772, 1773....... }	5	7	12	
Period of Immunity...	1774 to 1792 inclusive......				19
6th Epidemic Period..	{ *1793, 1794, 1795, 1796, 1797, 1798, 1799, 1800, 1801, 1802,* 1803, 1804, 1805, 1806, 1807, 1808, 1809...... }	13	4	17	
{ Period of Immunity..	1810 to 1817 inclusive......				8
7th Epidemic Period..	1818 to 1829 inclusive......	12		12	
Period of Immunity..	1830 to 1837 inclusive......				8
8th Epidemic Period..	1838 to 1844 inclusive......	7		7	
19th CENTURY, Period of Immunity..	1845 to 1850 inclusive......				6
9th Epidemic Period..	1851 to 1858 inclusive......	8		8	
Period of Immunity..	1859 to 1868 inclusive......				10
10th Epidemic Period..	*1869, 1870, 1871, 1872, 1873.*	1	4	5	
	GRAND TOTALS (extending over two centuries)..	63	42	105	98

trovertible certainty, that yellow fever never did, and therefore never will, appear as a land disease or a disease of the land, autochthonous or indigenous, on any of the West Indian islands.

GUADELOUPE.[*]

This French island, in latitude 15° north, was discovered in 1493 and occupied in 1635, when it was first attacked with yellow fever, as it was in 1640, 1648, and 1653. The intervening years were healthy. We then have to jump one hundred and forty years, viz., to 1793, for an outbreak, and then nine years to 1802. The malady prevailed in 1803, 1805, 1807, 1814, 1816, 1824, 1826, 1838, 1841, 1842, 1852, 1853, 1854, 1855, and 1856; so that up to the last year given by Cornilliac in his "Chronology," viz., 1861, there were only sixteen epidemic years and forty-five of absolute immunity.

Other Islands.

Let us take the islands on which it has been asserted that yellow fever was first seen:

Isabella, in latitude 19° north, discovered by Columbus in 1493, was afflicted in 1494 and 1495, and not once since.

Vega Royale, in latitude 19° north, is supposed to have been visited by the disease in 1496, though it was not regularly settled till 1500. Since then no disease.

Hayti, or San Domingo, situated in latitude 18° north, was peopled in 1498. It suffered three outbreaks in the sixteenth century, viz., in 1503, 1533, and 1585. Then we have to skip to 1792, 1801, 1802, and 1805; so that we have four invasions and only four years' prevalence as an epidemic in nearly two centuries. How does this indicate an autochthonous disease springing from the soil?

Trinidad, much nearer the tropics than the foregoing, viz., in latitude 10° 40′ north, discovered in 1498 and settled in 1797, has had one outbreak, in 1838.

Curaçoa, in latitude 11° north, settled in 1634, was affected with yellow fever in 1750 and 1760.

---

[*] "Recherches Chronologiques et Historiques sur la Fièvre Jaune," par J. J. J. Cornilliac. Fort de France, 1867.

Tobago, situated in latitude 11° north, has had two epidemics, one lasting two years, the other one.

Barbadoes, in latitude 12° north, settled in 1646, suffered in 1647, 1691, 1693, 1694, 1696, 1699, 1701, 1715, 1721, 1723, 1733, 1750, 1760, 1763, 1766, 1767, 1792, 1793, 1794, 1795, 1811, and 1814. This ugly-looking list only gives twenty-two years out of two hundred and fifteen, counting up to 1861. How did local conditions vanish which developed the disease only one year in ten, with two intervals as long as forty-four and forty-seven years?

Saint Vincent, in latitude 13° north, has had the disease once since its settlement in 1660.

Saint Lucia, in latitude 13° north, three times since 1639.

Dominica, in latitude 15° north, five times since 1660.

Montserrat, discovered by Columbus in 1493, has had four outbreaks since 1632.

Sainte Croix, in latitude 17° north, has since 1640 suffered ten years, of which six were in succession, from 1794 to 1800.

Saint Christopher is one of the islands early recorded. It is situated in latitude 17° north, was settled in 1625, and suffered in 1648, 1652, and 1653. During the present century it was infected, probably by the French fleet, in 1812.

Nièves, in latitude 17° north, established in 1628, has had one outbreak, in 1706.

Antigua, also in latitude 17° north, settled in 1629, has had but ten epidemic outbreaks, and the longest duration of the disease on the island was three years.

Porto Rico, in latitude 18° north, settled in 1508, was that year affected, and again invaded in 1843.

Jamaica, reputed not long since to be always infected, was discovered in 1494, in latitude 18° north. It was settled in 1509, and had no yellow fever till 1691—that is, for one hundred and eighty-two years after its being first peopled. Since then, and prior to 1861, it had only eight epidemics, the last in 1819.

Saint Thomas, in latitude 18° north, has had only three epidemics since 1650, the date of its settlement.

Port de Paix, in latitude 19° north, has had only one epidemic outbreak, lasting the two years 1690 and 1691, since 1663.

Cap Français, in latitude 19° north, had 17 epidemic years
in the 18th century, none in the 17th, and only two—1801 and
1802—in the 19th.

Port-au-Prince, in latitude 19° north, settled in 1668, had
the disease in 1795 and 1796. In the 19th century, up to 1861,
it was afflicted only in 1838, 1840, 1841, 1843, 1844, 1845, 1850,
1853, and 1854.

Lastly, Cuba, in 23° north latitude, settled in 1511, had no
yellow fever till 1762; then in 1793, 1794, 1800, 1801, 1804,
1811, 1819, 1829, 1833, 1838, 1840, 1844, 1861. This, "the
Pearl of the Antilles," whence the United States has been so
often infected, has had thirteen epidemic outbreaks in 350
years.

The foregoing data are certainly sufficiently accurate to show
that, on the most frequently, as well as the least frequently, affected of the West Indian Islands, the record of positive epidemics shows habitual health, and not a standing state of yellow-fever
contamination. It is, I believe, safe to infer that for the past
two centuries no single year has been free from infected ships
on the Atlantic. How and where did they become infected ?
Often, we know, they have shown disease in harbor without
communicating it to land, and it is very likely that, in a port
like Havana, with stagnant water contaminated by sewage, frequent outbreaks on ships have arisen while at anchor. This
gives great importance to Dr. Vanderpoel's suggestion respecting the purification of Havana harbor, and the project can only
result in good. I trust I have said enough in the foregoing
pages to indicate that a careful survey and sanitary history of
each island should be made, so as to prove whether the proxi-
mate truth we have so far gleaned is not confirmed by the
closest investigation.

I do not hesitate to say that none of the West Indian Islands—not one per se—can be reckoned as supplying the world
with yellow-fever poison. Germs or no germs, yellow fever is
not produced there endemically, and the secret of its development has been long locked 'tween decks, and in the fetid bilge.

CHAPTER III.

NATURE is sparing in means, but fruitful in ends. This trite lesson was early instilled into my mind by my great teacher in physiology, Dr. Sharpey. His learning and sound judgment influenced my labors, as they did those of all his pupils, more than the teachings of perhaps any other of his many colleagues. In part this was doubtless due to the nature of his subject, and to the paramount importance of a thorough knowledge of the functions of the body to the student of pathology.

Not in vain, I trust, have I observed that the very difficult investigation of disease causation is materially simplified by the recognition of the fact that the fundamental origin of indigenous and of contagious maladies is usually simple and singular. We are very apt to blur and hide it by publishing a catalogue of predispositions. Involved sentences and wordy generalities are made to stand in the place of fact and precise statement of definite knowledge. The practice of sifting and classifying actual data, rather than filling volumes with pleas and opinions, leads us by the only hopeful path. Beyond the truths clearly recorded and ascertained is that vast forest of briers and thorns, well calculated to entrap the dreamy thinker.

It is true that if we study the history of goitre and cretinism, of the Andean verrugas (an endemic of bleeding warts), of the *cachexia ossifraga* (fragility of bones) in animals in the hills of Lanarkshire and Central Europe, of the typhoid or enteric fever of large cities, we find that there are many unknown ele-

8

ments in the ætiology of each. But that is no justification for mystifying the essentials, or the known, by inferences which stretch to the unknown ; and I hope the time has arrived for a very careful culling of positive data from the mountain of chaff encumbering yellow-fever literature.

While writing these words, a friend brings me a lecture by Dr. Alfred Stillé, Professor of the Theory and Practice of Medicine in Philadelphia. Relating to the confusion caused by attempting explanations not warranted by facts, he says : " In default of any demonstrable or real cause, the usual refuge of ignorance has been eagerly sought for by theorists who are not content to seem ignorant of anything. They attempt to blind themselves and us with a cloud of words which describe or define nothing, and which, when reduced to their simplest expression, read ' zymotic poison.' Upon calm reflection this phrase turns out to be little else than ' words without knowledge.' "

As a sample of empty verbiage I may quote Dr. Griesenger as translated by Lemaître. He says : " Yellow fever is developed under the influence of general climacteric conditions : it is essentially a disease of the Western Hemisphere and of the Continent of America." Is it possible to compress within a round phrase greater inaccuracy ?

Very early in the medical records of yellow fever do we trace the thoughtful and judicious utterances of competent observers. Note the ring of the words of men who knew and spoke with prescience.

Sir Gilbert Blane, in a letter (dated 1798) to Rufus King, the ambassador from America to the Court of England, says : " It has been alleged by some authors that the yellow fever is produced by the same marshy exhalations which produce the intermittent and remittent fevers, and that it is only a variety of the latter ; but the remitting fevers differ from it in some essential symptoms, and the yellow fever has been known to arise both in ships and on shore where men were entirely out of the reach of the vapors of the marshes."

Dr. Joseph Bailey, Medical Officer of the Port of New York, writing in 1821, says that bilious remittents, remittents, and intermittents have nothing to do with yellow fever. " Does the principle of transmutability obtain in yellow fever? If it does, I have never been favored with a view of such transmutations. Hard or black frost has been long observed to destroy the contagion of yellow fever, but not to put an end to bilious."

When, in 1822, all were seeking, and, as they thought, *finding*, explanations of the local origin of yellow fever in New York and Philadelphia, Dr. Tully, of Middletown, Connecticut, defines its occurrence in small and healthy ports of New England, on the Connecticut River, as affording "the most incontrovertible evidence of its foreign origin"; and he adds: "There has been scarcely a season for the last five-and-twenty years in which individual instances have not occurred on the river, though the greatest number that has happened at one time has been at Hartford, at Middletown, and at Knowles's Landing; and, with the exception of a few theorists, there never has been a doubt of its foreign origin." And what Dr. Tully said of Connecticut we now know to be true of the whole United States.

At the very commencement of this century, Sir William (then Dr.) Pym showed that the disease is totally distinct from bilious remittent fever of warm climates, has no connection with or relation to marsh miasmata, and occurs in the West Indies (where he had seen what was called, after Chisholm, the *bulam* fever) under peculiar circumstances. Sir William was the first, or certainly one of the first, who demonstrated the immunity enjoyed after one attack of yellow fever.

Another sound practical man, Dr. Daniel L. M. Peixotto, of New York, formerly of Curaçoa, says, in 1822, "that it should be cited in conjunction with those places in the West Indies whose localities, abstractly considered, render them permanently unhealthy, neither experience will warrant nor my regard for truth suffer me to pass by without contradiction. The yellow fever is confined to the neighborhood of the sea; in the inland districts of the large islands and of the American continent it is unknown. There is a sufficient number of facts on record to prove that the fever occurs in hot latitudes at sea, *before new comers* have approached the land." *

Dr. Cyrus Perkins, of New York, addressed a very able

* "New York Medical and Physiological Journal," vol. i., 1822.

letter to Dr. Daniel Osgood, of Havana, in 1820, giving precision to the word *contagion*, as only applicable to a diseased animal secretion, and striking a broad line of demarkation between this and *infection*, or aërial poisoning by *matter* and *things*, not to be confounded with a specifically morbid animal secretion.

Dr. Stillé remarks:

"The late Dr. Nott, who spent nearly all his professional life in Mobile, and whose competence in such a question no one will doubt, states his judgment thus: 'Yellow fever is not generated in the human system, *nor transmitted from one person to another in any way;* its germ or poison is generated outside of the human system, and is taken into the system after the manner of the marsh-malaria poison. But, unlike the latter, its germ is portable, and may be carried from one point to another, and thus propagated.' And again he says: 'Few of the old and experienced physicians of the yellow-fever zone believe in the contagiousness of the disease, and their convictions are based upon facts coming under their observation. During thirty years' residence in Mobile, my experience corresponded with theirs.'

"The late Dr. Warren Stone, of New Orleans, who probably had more experience of yellow fever than any man who ever lived, stated emphatically the exact truth when he declared, 'I am perfectly convinced, beyond all doubt or hesitation, that, personally, it is not contagious; I know that it is not. In this city, at various times during nearly a century, local epidemics of yellow fever have occurred from time to time, every one of which was distinctly traceable to vessels from infected ports. Many of the patients were received into our ordinary hospitals, and perhaps not always with due care to leave behind their infected clothing; and yet in no single instance has the disease attacked their attendants or the surrounding hospital patients.'"

A striking contrast is furnished by Dr. Turner to "the average opinion of medical officers of the navy, who have observed yellow fever and experienced an epidemic visitation." The average opinion is as follows: "*In addition* to the causes of *malarial* fever, originating from vegetable decomposition, there is superadded a *miasmatic poison* from animal decomposition—*a fæcal poison.*" Now read the other, from an eminent member of Dr. Turner's own corps, who thus tersely and significantly puts the conditions: "Crowd filthy, half-fed emigrants in a filthy, unventilated ship, to cross the North Atlantic —what follows? Ship-fever, genuine typhus. Try the experiment in a tropical river" [he should have said the tropical

Atlantic—J. G.], " and you get *typhus icterodes* (yellow fever)."
I count the eminent member of Dr. Turner's own corps among
my supporters, in the opinion that yellow fever originates in the
ship.

However tempting may be a prolonged demonstration, I
must pause, not to unduly swell this volume. The most positive statement of the fundamental evidence, proving the distinction between the consequences of indigenous land miasm
and yellow fever, as also between relapsing fever and this last, is
furnished by three of the most learned and recent authors—Dr.
Faget, of New Orleans, Dr. Charles Murchison, of London, and
Dr. L. J. B. Bérenger Féraud, of Paris.*

Dr. Murchison † says that the frequency with which *relapsing fever* is complicated with jaundice has caused it to be mis-
taken for *true yellow fever*. The differences are stated very
distinctly under the following heads, omitting the last line,
which declares that "relapses of any sort are rare in yellow
fever," since it is universally reported that a patient appears
convalescent, leaves his room and house, and a deadly relapse
frequently supervenes :

a. Yellow fever exhibits no predilection for the poor and
destitute, but attacks all classes alike. Indeed, according to
some writers, feebleness of constitution prevents rather than
favors an attack.

b. Yellow fever attacks the same individual only once ; relapsing fever confers no immunity from subsequent attacks.

c. Jaundice is an almost constant symptom in yellow fever,
whereas it is much oftener absent than present in relapsing
fever.

d. Yellow fever is a most mortal disease ; relapsing fever is
rarely fatal.

e. Death in yellow fever is usually preceded by "black
vomit," which in relapsing fever, even when fatal, is so rare
that some of the most experienced of observers have doubted
its occurrence.

f. Lastly, the yellow fever of the tropics never follows the

---

* " De la Fièvre Jaune au Sénégal," Paris, 1874.

† " A Treatise on the Continued Fevers of Great Britain," by Charles Murchison, M. D., LL. D., F. R. S. Second edition, London, 1873.

peculiar course of relapsing fever—a febrile paroxysm, lasting
for a week, terminating in a critical sweat, followed by a complete intermission of a week, and then by a second paroxysm.

Dr. Féraud says: "Yellow fever *can not* be confounded
with the diseases of warm countries, except the bilious remittent (melanuric) fever." And here I translate his tabular exposition of symptoms, in parallel columns :

DIFFERENTIAL DIAGNOSIS BETWEEN BILIOUS REMITTENT FEVER
AND YELLOW FEVER.

Bilious Remittent Fever.	*Yellow Fever.*
Prolonged stay in a marshy country is the most potent and indisputable predisposing cause.	Prolonged sojourn in hot countries, whether or not marshy, gives immunity relative to the duration of such sojourn.
The disease is always preceded by many and frequent attacks of paludal fever; simple at first, then more or less complicated, and assuming generally more and more the bilious aspect, there being very notably anæmia of the subject.	The disease commences very generally amid a perfect health, and may occur in subjects who have never had intermittent fever, or who present the features of a most satisfactory state of plethora.
The malady is generally ushered in by a violent shivering fit, of greater or less duration, in every case similar to paludal fever.	The disease very frequently commences with a headache, which increases, and the commencement, which is instantaneous, can not be defined so well as the ushering in of an attack of paludal fever.
Jaundice appears forthwith with the first attack at the beginning of the malady. It is never absent, and causes from the beginning to the end a yellow color, varying from green-yellow to very marked ochre. The color is general and of the same tint throughout.	Jaundice appears consecutively about the third day, and takes the place of a red color of the integuments, which exists at the outbreak of the disease. The symptom is wanting in slight cases or when recovery is rapid. It is sometimes limited to certain regions, or presents some notable differences of intensity in different parts of the same individual.
The course of the disease is from the first intermittent or remittent,	The course of the disease is continuous from the first, and inflamma-

and the pulse, urine, and vomitings follow these changes very exactly. When the fever ceases, the period of weakness and reparation in no way resembles the remission of yellow fever, and is not separated, in a perfect and absolutely defined manner, from the first seizure; one might say that the fever yields unwillingly and attempting to return, if the patient succumbs in the febrile stage. If the patient reaches the adynamic stage, he dies more from a profound exhaustion than from the effects of decomposition.

The pulse follows the usual variations of the paludal fever, during a febrile period of two or three exacerbations, which constitute the first part of the malady. It does not fall suddenly, and is in this absolutely and in all like the pulse of a case of intermittent. Even when all is going favorably, daily observations indicate differences which constitute the vestiges of abortive attacks, so to speak.

Headache occurs like a load on the cranium, and increases for six or eight hours, then notably diminishes and disappears, sometimes to recur with the next attack.

The expression of the face is sunken, yellowish from the first or soon after the invasion of the disease. The conjunctivæ are of a yellowish color, but injected and bright as in a case of incipient conjunctivitis.

The pains of the body are round the waist from the loins to the hypochondriac region; great pain is

tory during two, three, or four days; one transition then supervenes, which is sufficiently marked to have deserved the name *mieux de la mort* (which may be translated *the calm before death*). For from six to twenty-four hours one may imagine the disease has ceased and the patient is becoming convalescent. The second period is perfectly separated from the first by this transition; it is a period of demolition of the subject, killing him by decomposition, suppuration, hæmorrhages, etc.

The pulse is at first full, regular as in a continuous fever, and it remains so until the transition termed *mieux de la mort*, or calm of death. At this period it sinks suddenly, and remains soft, slow, and compressible.

Supra-orbital headache is very intense, but it yields rapidly to remedial measures, and continues without intermission to the end of the inflammatory stage in from one to two days.

It is only after some days that the patient appears yellowish about the *alæ nasi*, the lids and lips. The eyes are bright, conjunctivæ injected, sometimes slightly bleared, as in a commencing conjunctivitis.

The lumbar pains which are called *coup de barre* are characteristic, from their intensity; they are very vio-

sometimes evinced in the region of the liver and epigastrium, and touching them causes lancinating pains, which cause cries; but often they are scarcely marked, so much so that sometimes these pains and the pains in the limbs present neither great acuteness nor great persistence, and are more like a feeling of uneasiness than marked pains.

lent, and do not extend round the waist. The hepatic and epigastric regions are not painful to the touch. There are frequently acute pains of the limbs, and especially of the calves of the legs.

The vomitings are bilious, of a very marked green color, most frequently like spinach water; they appear from the beginning of the attack, and cease with it, to return with the next.

Vomitings not frequent at first, and in all cases not bilious; they do not, moreover, manifest intermittence, as in bilious remittent fever.

After the first or febrile period of the disease, the vomitings continued, retaining exactly the same characters. They discolor linen of a lightish green, and if collected in a basin are very transparent, of a fine emerald green, or of an olive-like color.

After the inflammatory period, the vomitings when they appear are at first watery and colorless, then gray, then brown, then containing black matter like soot, giving a blackish color and not a limpid green to linen. Absolutely opaque when collected in a basin.

There is sometimes a bilious diarrhœa, from the beginning of the malady and during the vomitings. Later there is frequently diminution of stools, and one must have recourse to mild purgatives in order to keep the bowels in order.

Constipation very generally occurs at first; diarrhœa only comes on when the disease is prolonged, and then it is not bilious, but on the contrary very fetid, indicating a profound decomposition, and often containing the black matter, which is absolutely unknown in bilious remittent.

The tongue is moist, large, covered at first with a whitish mucus. This mucus soon acquires the greenish color of the vomitings. The tongue is not red, either at its top or at its sides; it remains large, swollen, and humid to the end of the disease.

Tongue white on its center, where it is furred, red at its top and borders; not so large, and somewhat globular. Later it is bleeding and shriveled (raccornie), of a mottled red as in typhoid affections.

The urine is black from the commencement, and the color is so characteristic that the patient is amazed

The urine at the outset is reddish and simply febrile, limpid, scanty, and discharged at rare intervals.

at it. It is generally very abundant and frequently discharged, and has only the melanic aspect during the exacerbations. Later the urine is still strongly colored, but no longer black. It sometimes contains a little bile at this stage—never at the beginning. Sometimes it is not abundant, but never suppressed except a few hours before death.

Later, if the disease becomes aggravated, the urine is thick and turbid; generally it becomes more and more scanty, and often at the last there is complete suppression for one or two days before death.

[It is albuminous.—J. G.]

The attacks at the outset may be arrested by quinine, and never demand antiphlogistics.

The fever continues from the start, can not be stopped by quinine, and often requires antiphlogistics.

The malady is manifestly linked with paludism; it follows and is followed by attacks of intermittent fever. It is absolutely non-transmissible from man to man.

The influence of paludism has not been brought out in an incontrovertible manner. The malady is not necessarily nor (may it be said) normally preceded or followed by attacks of intermittent fever.

The exacerbations are very frequent, and more and more ready as the attacks increase in number.

The return is extremely rare—so much so, that the possibility of a second attack has been denied by physicians.

Swelling of the parotid glands is very rare and quite accidental from the prolonged administration of calomel determining stomatitis. I have only seen two or three cases of this kind in over three hundred observations, and the relation of cause and effect was readily established in each case.

There is frequent swelling of the parotids toward the close of the malady.

"It is obvious," adds Dr. Féraud, "that the differences are so sharply defined that I hope hereafter no one can for a moment confound the two diseases. After this comparison, hesitation, which comes from the assertion of facts which have been superficially or insufficiently observed, will finally disappear, to the great advantage of the sick." Dr. Féraud states as a character of yellow fever that its transmission from man to man is terribly frequent, but this is almost universally denied by the best authorities.

Dr. Faget, of course, points to the origin of yellow fever, not only on the ocean, but in the ships—"une origine non seulement maritime, mais navale." That malarial fevers pertain to marshes, and only to marshes. The air of the infected ship, like the air of the marsh, may travel within certain limits, and winds have thus tended to propagate both. Yellow fever is transmitted—marsh fever never. The marsh fever, which can be wafted by the winds, probably farther than yellow fever, can not, as Mélier puts it, be charged or loaded on board ship. For more obvious reasons it can not be carried in a box, in clothing, and especially woolen garments. It is the *mediate* communication which gives to yellow fever the aspects of contagion, especially when the things which carry it are on the bodies of the sick. The cases to prove contagion demanded by Chervin (" Chervin était difficile ; c'était son devoir de l'être," says Faget) " were cases in which the patient had been divested of all clothing, especially of woolens coming from the place where the disease had originated."

" Paludal fever continues after frost," says Dr. Faget, " but I have never seen yellow fever after simple white frost."

And lastly, he says, so far as yellow fever is concerned, the morbid principle, the poison, can be destroyed, whatever it may be, before there is any contact with susceptible individuals. For this it is *sufficient* to purify, in an isolated locality, the ships which import it, which carry it especially in their holds after unloading. To be more certain, it is prudent to exercise some precautions with regard to merchandise and also with passengers.

In marsh fevers the poison can only be traced, reached, and killed in the system by the use of the one specific which art has given us, viz., quinine.

It is to be hoped that henceforth all those who claim an adequate medical knowledge to report on cases and outbreaks of yellow fever may cease to air their theories on marsh and malarial poison, " epidemic influence," and " epidemic constitution." I have had many a battle with these mouthy *impedimenta*, whose authority in my country outweighed for long all industry, research, and well-founded perception.

The Pulse and Thermometer in Yellow Fever.

Dr. J. C. Faget* has cleared away many doubts and difficulties as to the nature and diagnosis of yellow fever, by his painstaking and enlightened researches during epidemics in the South. In determining the continuity of the febrile action of yellow fever, hence its specific distinction from paludal fevers, he says : " If there be in the world a marshy region par excellence, it is New Orleans ; and if there be, consequently, in spite of statements to the contrary, a locality rich in marsh fevers of all sorts, which very frequently complicate the maladies and communicate to them their own paroxysmal features, it is assuredly our unfortunate city. If, then, in such a region, the symptoms of yellow fever, as observed by us, are found to be those of *continued fevers*, or of fevers having *a single paroxysm*, it will be proved that yellow fever is everywhere, essentially, a fever of the continued type." Furthermore : " The four or five great epidemics of yellow fever which we have experienced at New Orleans during these last twenty years, from 1853 to 1873, were all complicated with malarial fevers, of the hæmatemesic (black vomit) or hæmagastric type ; hence all our difficulties."

In yellow fever only *one attack, one paroxysm, one access*, and not two, never a true remission, except the final, and this *single access*, very remarkably, declines from the time it commences ; its course is *descendent* almost immediately after it manifests itself. The duration of the pyretic action is six to seven days. The temperature rises rapidly to ordinary fever heat, 40° Cent., or 104° to 105° Fahr. ; this is termed the period of effervescence. This short and rapid period of increase is followed by a tolerably slow and protracted period of decline—that of defervescence.

In ordinary fevers the fall of the temperature is more rapid than its rise. It is the contrary in yellow fever ; in it the effervescence continues from one to three days, and the defervescence from four to seven days ; this gives *twelve days* as the total mean duration of the febrile temperature in yellow fever ; whereas, in the initial fever of variola and varioloid, which has

* "The Type and Specificity of Yellow Fever established with the aid of the Watch and Thermometer." Paris and New Orleans. 1875.

so much resemblance, at the commencement, to yellow fever, the *rise* of temperature, or effervescence, is four days, and the defervescence two days.

In the fatal cases, the defervescence, arrested in its descent by the visceral congestions which we have described, in those cases, sometimes assumes an *ascending course toward the end till death*, but *not always*. Sometimes the temperature continues *to sink during the death agony*.

There is a rapid and early rise of the pulse up to one hundred and twenty pulsations or almost double the normal figure; then the pulse sinks in frequency slowly and with increased slowness as the disease progresses. The maximum average being 120 the first day, the second it is 10 or 12 beats less, then in 24 hours a dozen fewer; on the fourth day it oscillates toward 80; it reaches 60 the seventh or eighth day; on the tenth an astounding *minimum* will be occasionally obtained; in some patients, in convalescence, the pulse does not exceed 40 in the minute. There is a little habitual *evening exacerbation*, which the lines of temperature show in yellow fever, as in almost all fevers. The tendency of the pulse to lessen the number of its pulsations does not, in yellow fever, proceed to the extent of rendering the heart insensible to the effects of secondary visceral congestions. When the termination is to be fatal the pulse rises, exceeding the early maximum, and being at last so that it can not be counted, but with this terminal rise of the pulse there is a fall of the temperature.

The fever-producing principle is a concrete entity, material and real; it escapes the senses, because it is infinitely minute, but *the effects* which it produces are appreciable by the senses, and these are the specific effects which enable us to assert its real and specific existence. It is an entity invisible, impalpable, inaccessible to our senses; but it is so real that it *can be destroyed* where it is known to exist, as in the *holds of certain ships*, and sometimes in ships on which there are no sick people, or on which, at least, there have been no sick for a long time. This yellow-fever-producing principle is quite apart from all human organisms, and existing by itself, whatever it may be.

Of all fevers, yellow fever alone presents a steady diminution *in the number* of arterial pulsations from the onset, directly the *maximum* of the pulse is attained, that is to say, from the initial hours of the fever, at the maximum of the febrile reaction of the organism; this, with the rising temperature, indicates a probable *special action* of the fever-producing principle on the heart.

We therefore conclude, according to Dr. J. C. Faget and in his very words, that in yellow fever, from the first, the line of the pulse descends, while that of the temperature maintains itself horizontal in the great majority of cases, or even rises for two or three days or more, in two thirds of the cases at least. This is the gnomonic clinical symptom of yellow fever.

The continuity of the pyretic action, with divergence of the lines at the commencement, the line of the temperature in particular maintaining itself high, ought to make the balance incline to the side of yellow fever. On the other hand, digressions, *abrupt falls*, especially in the temperature, with *sudden risings*, involving *parallel elevations* of the line of the pulse, will reveal to the physician that the paludal element is at work, and will induce him to have recourse boldly to quinine, from the commencement of these hæmorrhagic fevers of ataxic form, which have a course equally deceptive and full of peril.

In relation to *prognosis*, the two lines *descending parallel*, at least from the third or fourth, slowly and steadily descending, is the most assuring fact in the course of yellow fever. If, in their descent, they stop and present *horizontal oscillations*, something abnormal has occurred; then is the time to subject the organs to an attentive examination, and to *reserve the diagnosis*.

Lastly, a little later, if the two lines show a *divergence inverse to that at the commencement*, if that of the pulse takes an ascending course at the end, while that of the temperature sinks low, death is almost certain, or, rather, it is even impending.

The name yellow fever is derived from that special symptom which may not appear in *all* the sick but in all it destroys. It is quite a special kind of persistent yellow, not due to jaundice, in which nitric acid at once detects *biliverdine*, but, with the yellowness, albumen appears in the urine. The *black vomit* is very marked, and perhaps not a single body, dead from yellow fever, but has the stomach full of it. The true black vomit has the appearance of the infusion and grounds of coffee. Corpses in other fevers may contain it.

Acute fatty degeneration of the liver seems also to be a structural alteration constant after yellow fever. Like black vomit, it is found in the bodies of those who have died from other fevers; but, when it is not found in a case believed to have been yellow fever, it is probable that there has been an error in diagnosis.

Such are Dr. Faget's views, and they may be regarded as the most recent and definite exposition of genuine originality.

In a highly interesting letter on " The Epidemic Fever of Young Children now prevailing in New Orleans," which he addressed to the New Orleans " Daily Picayune," on August 24, 1878, Dr. Faget shows the vast importance of the diagnostic signs he has discovered, and of the consequent adaptations of treatment to the recognized character of the disease. He says :

" Yellow fever has been rapidly spreading in our midst. But, at this season of the year, in August, near to September, is our city and is the country around it exempt and free from malarial fevers, and have not these fevers a larger share of the mortality than is allowed ?

" Can not both the yellow and the malarial fevers coexist in the same locality, and do they not sometimes attack simultaneously the same person ?

" Is it positively to yellow fever that are due all the deaths of young children of only a few months, or even of a few years (two, three, four years . . . .), which appear every morning in the papers in the list of deaths from yellow fever ?

" An important and well-established fact is, that the older the patient the more severe is the type of yellow fever. Under ten years, dangerous and particularly fatal cases are very rare.

.   .   .   .

" It is an undeniable fact that in this city, during the first fifty years of this century, that is from the first appearance of yellow fever here, in 1796, until 1853, it was considered that our children, born in the city, had nothing to fear from the disease ; it did not seem that they had ever been affected by it.

" But in 1853, and even more so in 1858, while yellow fever was decimating the unacclimated foreigners, a great number of young children, born and raised here, were carried away by a fever presenting some analogy with yellow fever. Many physicians believed that it was yellow fever, and abandoned their former opinions on the subject. I am one of those who still adhere to the opinion that the natives of this city are exempt from yellow fever; that they are not subject to it any more now than formerly.

"Already during the great '*double*' epidemic of 1853, in an attentive study of the course of the disease which was striking down the children, some of the cases being with 'black vomit,' I had observed first, that the course of the disease was similar to that of pseudo-continuous paludal fevers; and second, that the disease yielded to sulphate of quinine, when given in time, and in sufficiently large doses.

. . . .

"1. The febrile action or force in this fever of children manifests itself by '*exacerbations*,' while in yellow fever there is but one paroxysm.

"2. The matters *black-vomited*, instead of resembling an infusion of coffee, with the grounds at the bottom, of the yellow fever cases, generally presented *black grumes* with a great deal of mucus from the stomach, heavier than water, so that, in order to find it at the bottom, under the vomited liquids, they had first to be carefully poured out. The liquids are of varying brown shades, reaching the black color of prune juice.

"3. The black vomit in the paludal fevers is not a sign of much gravity, while in yellow fever it is almost a sure sign of approaching death.

. . . .

"4. Except in cases complicated with jaundice the corpses were' not yellow as they always are in yellow fever, even when the yellow color did not appear during life.

"5. The relapses are as frequent as they are rare in yellow fever.

"6. Quinine, which is of doubtful efficiency in yellow fever, was the 'specific' remedy in this fever of children. In 1858, in the epidemic of the children, where quinine was not used, the proportion of deaths was as great as in yellow fever; and, on the contrary, where quinine was given in sufficiently large doses, and when it was absorbed in the system, there were but few deaths.

"It strikes me that those differences are sufficiently marked and real to justify the opinion, that the fever, which decimated the young children in New Orleans, during the yellow fever epidemic of 1858, was not yellow fever.

"In the last mortuary report for the week ending Aug. 18th, we find 315 deaths within the limits of the city of New Orleans, of which 185 deaths are imputed to yellow fever. The mortality, divided according to age, is as follows: Under 1 year 20, from 1 to 2 years 22, from 2 to 5 years 50, from 5 to 10 years 16.

"It would thus appear that the more tender the age, the more deadly the yellow fever.

"This would be the complete subversion of all that was known hitherto of yellow fever, as studied in all countries, by all writers and at all epochs.

"Is nature no longer constant and invariable in its laws? Or is not this epidemic fever of young children, which is called 'yellow fever,' the same which prevailed in 1853 and 1858, during the yellow fever epidemics of those years, and of which I have given above, as before in 1859, the differential characteristics as compared with yellow fever.

. . . .

"If yellow fever, from the very beginning, presents almost constantly this characteristic of a diminution of arterial pulsations, without any decrease of the febrile heat, and sometimes even while the fever is increasing, if this sign belongs only to yellow fever, then we have at once the means, from the very first hours, to know it from among all other fevers. It furnishes a prompt and safe guide to decide whether the present fever of young children is or is not yellow fever.

"If, at the very height of the fever, and from the beginning, there should be observed, in the sick children, a diminution of the arterial pulsations, then the fever would be the yellow fever. This is what I have never found either in 1853, in 1858, or during the present year. On the contrary, in this fever of young children, which, in all cases coming within my observation, I have always found (unless quinine had intervened energetically from the outset) to be a fever with several paroxysms or exacerbations, there is no discordance or opposition between the pulse and the temperature; in other words, if the febrile heat increases the pulsations are more rapid.

"I hear it said rather vaguely: The progress of fevers is not the same in adults as in children. This may be. But let us be precise. Do not poisons generally produce, all proportions being allowed, the same effects in children as in adults? Do not digitalis and veratrum, causing a diminution of the arterial pulsations in the adults, produce a like diminution of pulsations in children? Why, then, should not the yellow-fever poison, which causes such a remarkable diminution of the pulse in adults, cause also the same diminution of the pulse in children?

"Therefore, if, in this fever of children which we are now considering, there is no diminution of the pulsations, while the temperature increases or remains stationary; if there is no discordance between the pulsations and the febrile heat; if both the pulse and the temperature increase during the paroxysms, then we say it is not the yellow fever.

. . . .

"The existence of this hæmatemesic form of the malarial hæmorrhagic fever can no longer be denied. It was even known in the days of Hippocrates, and is mentioned by him. It was also known by the physicians of Spain as early as the middle of the last century, long before yellow fever had been imported into Spain from its colonies. And although it was supposed by some, in 1869, to be a new disease, it had been mentioned and described since about twenty-five years by the physicians in the Antilles.

"We consider that the existence of the hæmatemesic form of the malarial hæmorrhagic fever is as clearly established as its hæmaturic form.

"My conclusion is that quinine should be freely given in the epidemic fever which now prevails on young children.

. . . .

" The only means of knowing positively that it has been absorbed is to ascertain that it has produced its physiological effect, the hardness of hearing.

" If quinine can not be given through the stomach or the rectum, frictions can be used, with some benefit, in cases of small children whose skin is very delicate. But in severe cases it is always unsafe to trust too much to frictions, and recourse should be had to subcutaneous or hypodermic injections; they sometimes cause local inflammation; but, again, what is that in such a danger? As Hippocrates so excellently says in his Aphorisms: ' To extreme ills, extreme remedies.'

" DR. CHARLES FAGET."

I have diligently sought for any evidence of the land origin of a disease similar to yellow fever, and perhaps the most striking on record is

Oroya Fever.

The recent construction of the Callao, Lima, and Oroya Railroad has been attended by the manifestation of a pestilential fever of great malignity, confined to the line of grading, which had no previous existence, and which is commonly called the " Oroya fever."

Dr. Browne, Fleet Surgeon of the Pacific Station, writing in 1872, says it commences " at La Chosica, 33 miles from Callao, at an elevation of 2,800 feet; its locality extends along the course of the road through the valley of the Rimac River for about 22 miles, to the elevation of 6,500 feet"—in the same locality where prevail the *verrugas* and simple intermittent fevers. Dr. Fasset, of Lima, considers it "but an aggravated form of the pernicious paludal intermittent fever that is common to marshy localities, or where rice is cultivated, and particularly in the deep, hot, and tumid valleys of the Sierra."

The atmosphere circulates badly in the humid gorges of the Sierra. The work upon the road in many places, by the ground being broken up, has occasioned the emanation of a fetid odor, more offensive than that of sulphuretted hydrogen. " It may be," says Dr. Browne, " that this telluric miasm, acting upon a system previously impressed with a miasm of the ordinary intermittent, may develop a highly malignant fever." It attacks the whites, the mongrel, and especially foreigners. Negroes, Chinese, and Indians are most exempt. It subjects to a second attack, does not preserve from yellow fever, nor does the latter

9

grant immunity from it. It is non-contagious, non-ambulant, or non-transmissible, and removal to a short distance is sufficient for withdrawal from its influence.

"Departing from Lima, the railroad follows the left bank of the Rimac, and until it reaches La Chosica the workmen are only subject to simple intermittent; from thence, where the verrugas waters commence, they are exposed to this pestilential fever up to the point where begins the cold climate of the Sierra."

Oroya fever is sometimes continuous, at others periodic. Its features have confounded the medical observers in all "save its lamentable frequence and fearful destruction." Dr. Browne says:

"Death may occur in twenty-four hours from the attack, though the duration of the disease usually is several days, and in some cases prolonged into weeks. Dr. Rush, an American physician in the company's service, died in thirty-six hours from the first invasion, apparently in ordinary health until the moment of the attack. In certain cases a slight attack is followed by a seeming convalescence; the patient goes out, a relapse ensues, and he dies suddenly. It is the rival of yellow fever, equally formidable, and, though differing in many of the symptoms, the result is nearly the same. It has been observed that, if the access belongs to the quotidian type, it is favorable if it delays; and, on the contrary, it is unfavorable if it anticipates when of the tertian variety.

"In the commencement the symptoms are usually the same: intense cold, accompanied by severe pain in the head, loins, and limbs, succeeded by febrile movement, varying in intensity, with continuance of headache, pain in loins, etc. This stage is followed by an intermission or remission, or the severity of the fever may subside into a feverish condition which has continuance, or the paroxysm may be succeeded by copious sweats, affording no exemption from other paroxysms. The intermissions and remissions are exceedingly regular, or after a single one the fever may become continuous. Prostration, relaxation of the muscles, and anæmia are often sudden, and followed by aphony.

"Early in the fever nausea and vomiting usually occur, the matter at first being yellowish and greenish in color, then brown, then resembling coffee-grounds, and finally melanic. There are severe pains in the region of the spleen, liver, and stomach; no tympanites when unattended with peritonitis, but contraction of the abdominal walls toward the dorsal spine; in some cases iliac tenderness, with absence of gurgling sound. The bowels are generally constipated, therein differing from the grave cases of the pernicious intermittent of the country, when diarrhœa, often bloody, supervenes. Intelligence may be good, or a maniacal delirium at the outset. Petechial spots occasionally found; urine brown, as in yellow fever, but not often suppressed, except with peritoneal complication. Tongue presents a gray coating or greenish yellow at first, but becomes red and raspatory. Sordes not frequent, gums bloody, breath tainted. The blood seems deprived of hæmatosine and globules. The patients take an icteric tint like that of yellow fever.

"As the disease progresses the ataxic symptoms denoting the typhoid state are evident, viz. : low delirium, sordes, subsultus tendinum, etc. The comatose or convulsive state frequently happens, and the last condition appears to be less dreaded, however violent it may be, than the first, except a nervous trembling of the limbs and tongue, a bad augury.

"I was unable to obtain any satisfactory information as to the morbid appearance after death, *post-mortem* examinations having been very rarely performed. Dr. Fasset asserts, however, that the pathological alterations produced by yellow fever are perhaps surpassed by the pernicious paludal intermittent fever, of which he regards the Oroya fever as a variety."

The manifest importance of tracing every form of disease on land bearing any similarity to yellow fever must be my sufficient excuse for so long an extract.*

Cholera and Yellow Fever. Contrast of a Land and an Ocean Plague.

Scarcely possible does it seem to me that the great disparities between a disease springing from the soil, like cholera, and an ocean pestilence, like yellow fever, should not long since have forced on medical men the view that the last is a disease of the mariner.

* On Oroya Fever in "Medical Essays by Medical Officers of the U. S. Navy." Washington, 1872.

Hindostan, guarded from the world beyond it by the Hima-
layas and a shore distant from European ports, protected by caste
prejudices, which have preserved ancient rites and a peculiar
civilization, retained till late many secrets, and among them a
knowledge of its virulent endemics. Its pestilential jungles and
still more pestilential salt marshes have for ages bred remittents
and dysenterics. Our knowledge of the people of India and
their diseases dates back only to Vasco de Gama, toward the
close of the 15th century. One of his countrymen, a Portuguese physician, Gaspar Correa, describes cholera in the " Lendas
da India " as killing in 1503 20,000 men in the army of the Za-
moryn sovereign of Calicut, the enemy of the King of Cochin.
Here and later at Goa (1543) the disease, unlike yellow fever, is
observed spreading among soldiers on land as it has done, moving from camp to camp, among the British forces since we have
held possession of the East Indies. Oola Beebee, the goddess
of cholera in Lower Bengal, specially worshiped during the more
deadly months of April, May, and June, still exists to bear testimony to the antiquity of this indigenous plague (Macnamara).

The annals of cholera teem with observations from 1503 to
1807, when it first became epidemic under British rule. When
Dr. Tytler first wrote of the disease on the 23d of August,
1817, he said : " The disease is the usual epidemic of this period
of the year," and the magistrate of Calcutta, writing to the
government in September, observed that a disease prevailed, as
it generally did to a greater or less degree " at the present season of the year "—" of the species of cholera morbus."

No such record is given, by the early and trustworthy American authorities, of yellow fever, which they believed had sprung
upon them, under some mysterious epidemic influence, in the
busiest seaports and nowhere else.

Of the six great Indian outbreaks of cholera,* the very first
one in 1817, within three months from its appearance, had been
generated *throughout* the province of Bengal, including some
195,935 square miles, and within this vast area, says Macnamara,
the inhabitants of hardly a single village or town had escaped
its deadly influence. In 16 months the disease was generated

---

* " A History of Asiatic Cholera," by C. Macnamara, F. C. U., Surgeon to the
Westminster Hospital, London, 1876.

throughout the length and breadth of Hindostan. Thus spreads a land plague.

When yellow fever entered Philadelphia in 1699, it killed 220 people out of a population of 3,800, and as usual was confined to the port. In the ever memorable outburst of 1793, in the same city, 3,548 died out of a maximum number of 50,000 inhabitants, and the one third of these who escaped inland propagated no disease. I am aware that I can quote numerous instances of most grievous loss of life, such as in Spain at the beginning of the present century; but, whenever an outbreak of yellow fever is studied, it will be found dotting about among the seaport and river towns, and not besmearing, as you can a map of Hindostan with a paint brush, an entire continent within the limits of ocean and impassable mountains.

It is highly interesting and instructive to note the universal and invariable tendency of the human mind to discover an immediate cause for epidemic outbreaks. Just as the appearance of yellow fever from the very first developed the erroneous opinion of its origin in New York, Philadelphia, and elsewhere, but always locally, so when cholera first broke loose in 1819, and was taken to the Mauritius, the local commissioners assembled were unanimous in not supposing the disease contagious, or of foreign introduction; the cause was supposed to exist in the atmosphere.

Each and every year cholera occurs inland during the hotter months, and ceases in winter. The great epidemics like that of 1817-'23 wear away. Recrudescence each year occurs in all directions, and not, as in yellow fever, invariably near an expanse of water—ocean or river.

Yellow fever may be said to wear away; but, as innumerable observers have noted, and Drs. Faget and Hargis have stated, the seasonal remission, as a rule, demands renewed importation. You must resow it like the cotton plant, which lives perennially in the tropics, but can not resist the winters in the United States, and like the indigo in Tirhoot, which must be raised from imported seed.

When cholera spreads beyond Bengal—its true home—it spreads by pollution of water; and, if not renewed by fresh contact with Bengal, the disease in those parts where it has reached gradually dies out and disappears. Where *imported*, cholera seldom continues for more than three consecutive seasons in any one place.

Yellow fever passing from the ships to land may and does in hot countries remain for years in seaport towns, but the smaller of the West Indian Islands are free from the disease in their interior, and only retain the disease for limited periods. They all enjoy protracted periods of immunity when there is absolutely no sign of the disease. The remark that Dr. Blair makes concerning Demerara, which escaped from 1819 to 1837, shows this ; for, when the malady broke out in the latter year, no one could be sure of its nature, for the physicians who had seen the previous epidemic were either dead or had left the colony. The same would happen in the case of cholera beyond its home ; but there is no land, no island, and only a seaport in communication with shipping, where yellow fever can demand medical attention every succeeding year.

The appearance of cholera in America for the first time in 1830, '31 and '32, was due to an epidemic outburst which commenced in Bengal in 1826. It traveled through many lands, and crossed vast continents, as well as seas, to reach the New World *via* Dublin. As Macnamara says, this " mighty continent contains people living under every variety of climate and of varying social circumstances ; none of these, however, had developed cholera among her people."

Could yellow fever ever reach Bengal across continents ? Even ships fail to carry it there. It can not travel by easy stages from place to place, over months and years, and continue to exist, much less to destroy with pestilential virulence.

Cholera affords proof how persistent and troublesome an exotic land disease may be far from its well-known home. Writing in July, 1875, Mr. John Simon says : " Recent facts as to cholera in Europe have undoubtedly been of very evil omen. Europe within the ten years time has twice been overrun with cholera. From the middle of 1865, when one great diffusion of the disease began, till after the middle of 1874, when a second great diffusion had apparently run its course, there possibly was no moment at which the disease was extinct in Europe, and

there certainly was but little time when it could even be supposed to be extinct."*

We are now witnessing an exceptional persistence of yellow
fever in Memphis, but has it disseminated itself like a land disease? Did it do so even in 1878, that year of exceptional
invasions at points somewhat remote from water, though still
within easy reach of Mississippi and its infected steamers?

Yellow fever does not demand, as *conditio sine qua non* for
its development, the introduction of a sick human being into a
port to propagate the malady. Frequently the ship's company
has been landed, and the susceptible laborer, stevedore, custom-
house officer, or ship-keeper has been the first to contract the
malady, and with others aid in its transportation from the ship
to the port. An example of propagation, essentially non-per-
sonal, is afforded in Surgeon-General Woodsworth's tenth report, dated September 14, 1878. On the 4th of September
four members of a family, residing two miles from Cairo upon
the banks of the Mississippi, were suddenly stricken with yellow fever. On the 18th, the remaining two members of the
family were attacked, one died. The children found an abandoned skiff on the river, at a time when infected steamers had
been working mischief, and removed it to the house for repair.
Three or four days afterward all who were around the boat were
stricken with the fever in one day. The extension of cholera,
on the other hand, bears a close and definite relation to personal
traffic : in various important cases the arrival of persons affected
with the disease was unquestionably the starting-point of local
and perhaps national epidemics ; and no extension of the disease was to be found where the arrival of human beings from
previously infected places was not either proven or probable.
(Simon.)

Outside the limits of India, Mr. J. Netten Radcliffe, in his
very wide study of cholera, extending from England to the torrid climates of Africa and Southwestern Asia, finds no reason
to impute to cholera any other mode of origination and extension than such as that doctrine expresses.

* "Reports of the Medical Officer of the Privy Council." New Series. No. V.
London, 1875.

Analogy between Typhus and Yellow Fever in a Ship.

Dr. Mélier reports a highly instructive instance of ship typhus, which occurred on the imperial transport Duperré, employed in bringing back to France the wounded from the Crimea, which were in that state of debility and sickness named by Michel Lévy " Crimean cachexia." Having left Eupatoria on the 18th of April, she arrived at Toulon on the 2d of May. In the 22 days' voyage there were many deaths. The vessel was purified with scrupulous care—450 men remained on board and continued to sicken. On the 13th of May, viz., after 11 days, 23 cases had been sent to the Hospital Saint Mandrier, five or six very severe and presenting the symptoms of typhus. These, it might have been thought, were cases of prolonged incubation, but on the 22d new cases occurred. M. Mélier examined the vessel with great diligence. Everything seemed in perfect condition. Admiral Dubourdieu, then Maritime Prefect of Toulon, graphically pronounced the verdict, " *C'est le navire qui est malade* " (it is the ship that is sick). The time came for the men to leave the ship ; their time had expired. Another crew takes its place, all new men ; and they are seized slightly, it is true, but with unmistakable symptoms—demi-typhiques.

CHAPTER IV.

THE contrasts, causal and symptomatic, between different morbid conditions, embraced under the general head "fevers," suggest deeper differences, which may be defined as *developmental*. Acute observers have shown us the essential signs of demarkation between periodic and continued fevers—between eruptive fevers with typical structural lesions and fevers which are unattended with a specific anatomical change.

It is in my opinion impossible to exaggerate the importance of Dr. J. C. Faget's diagnostic investigations, in relation to paludal and yellow fevers. They must mark an era in the study of these diseases; and for the purposes of medical practitioners during epidemics, they constitute the true beacon lights, which should keep out of print a host of immature observations, capable of materially hindering the growth of exact knowledge. The looseness of records, and the cloud of baseless theories, are responsible for that chaotic complexity of medical opinion, which I desire, in some measure, to dispel.

The researches of great difficulty and importance into the ultimate nature of true contagion, and the primary causes of some diseases of local origin, have led us one step beyond the bedside observations. They have shown, in some cases, that special elements are at work, associated with and characterized by peculiar structural formations and living organisms. Many mere coincidences have been interpreted as cause and effect. It were well if all coincidences had been always noted to the fullest possible extent, and then the natural growth of biological knowledge would have explained phenomena beyond the ken and grasp of the prescriber. His primary duties, responsibilities, and fatigues unfit him for calm and continuous thought and work on the deeper and most difficult problems of life, while the battle with death is being fought red-handed. The first injunctions to the clinical student should be to note everything, however trivial, and never trust to memory. The accuracy of medical writings would be enhanced immeasurably if we could delete from past records the imaginative pictures of cases which have been drawn with rounded phrases and well-chosen periods, months and years after observation. They are stumbling-blocks in the way of the accurate philosopher. They are just as pernicious, in medicine, as fanciful and incorrect representations of animals and plants would be to the anatomist and botanist.

Hasty generalizations have materially retarded the development of medical knowledge, and I desire, in submitting to my readers some speculations based on the knowledge acquired, to be more suggestive than doctrinaire—thought-inspiring rather than limiting mental effort, even by reasonable explanations of manifold phenomena.

A morbid principle—a fever-poison—has been very widely accepted as a *blood-poison.* So common and constant is the dispersion of intoxicants through the system by the organs of circulation, that the instant and primary effect of an infection is interpreted as due to an action on the blood. The surgeon who talks of "shock" apprehends the all-pervading influence of the sensorium, and its subsidiary ganglia and nerves. The soldier who, as Guthrie relates, was killed by a bullet striking the pit of the stomach, producing only a contusion, affords an instance of that exquisite relation between animate man and his surroundings which has bewildered so many.

So hard is it to discipline the intelligence to any limitations that, by a bound, the metaphysical labyrinth is reached, even by our best experimentalists. No less a man than G. A. Hirn [*] has compared inorganic matter—homogeneous, susceptible of being broken into fragments, and of indefinite continuance in a

* "Conséquences Philosophiques et Métaphysiques de la Thermodynamique." Par G. A. Hirn. Paris, 1868.

state of absolute unchange unless disturbed by external influences—with the living being, essentially heterogeneous in structure, with diversified organs and functions necessary to limited existence. Internal rest of its constituent parts is in direct opposition to life; it grows and feeds on its surroundings, and the smallest instant of time, during which the necessary changes of its integral constituents may be stopped, suffices to kill and convert the living into ordinary matter of the physical world.

The living being creates nothing—it only assimilates. If parts or things are thrust on it, not compatible with its organism, it makes a supreme effort to reject them, and failing in this dies. " If, therefore," continues Hirn, and here I quote literally, " a special principle is necessary to the constitution of the whole (*l'ensemble*) of a living being, the principle is, once for always, in it, qualitatively and quantitatively, from the moment of its birth to that of its death. This principle can not be regarded as a force." . . . . " The expressions organic forces, vital forces, have no meaning." And he goes on to argue in favor of a *principe animique* as a necessity, that is to say, " as an affirmation which subsists because no other can be found to take its place." Where his demonstrations fail, he deems it imperative to impose a dogma.

The origin and propagation of yellow fever indicate the development of a virulent something which, within certain limits, reproduces the same morbid appearances in susceptible individuals. One of the most striking and distinctive of its features is that this virulent entity exerts such an influence on the human system, as to make this resist, more or less effectually, any subsequent attack.

And by way of parenthesis I must here note how some of the ardent disputants have attempted to bend facts, or rather to substitute assertion for fact, in support of their theories. In attempting to controvert the opinion that yellow fever was a contagious disease, and before Sir William Pym and others had demonstrated freedom from second attacks, Dr. Charles Maclean had " *completed his convictions* " that " no general disease which is capable of affecting the same person repeatedly is ever propagated by contagion "; *ergo*, after Pym's researches it was admissible to consider yellow fever a pure contagion. Experience lends much more weight to another dictum which reflects great credit on Dr. Maclean's perceptions, especially if we consider that he belongs practically to the last century. He laid down, perhaps not absolutely for the first time, the principle " that a disease capable of being propagated by a specific *virus* can never be produced by any other cause; and that a disease produced by other causes can never be propagated by a specific virus." *

But yellow fever is transmissible, though not inoculable, and, in all probability, always an imported disease on land, notwithstanding Maclean's logic and Chervin's eloquence; and it is necessary we should show in what it differs from the *pure contagia*, the specific inoculable diseases, with which, as I have before explained, it can not for one moment be confounded.

We must not be rash in our inferences, but bear in mind the advice given by Dr. Latham in his clinical lectures. He says : " Let a man use his own experience as best he can for the present, but let him not, upon the strength of it, rebuke the experience of all past times and dictate to the experience of all future ; for if he live long enough, nothing is more likely than that he may find himself under his own reproof, and inconveniently confronted by his own maxims."

In the distinction between inoculable and non-inoculable maladies, we may find cause for confusion of thought. Usually diseases dependent on a specific virus are inoculable, whereas other maladies are non-inoculable ; but just as a sound piece of meat may be inoculated by that ferment which causes putrefaction, so do we find, as I have personally witnessed and actually suffered from, non-specific venom communicated from irritated and inflamed mucous surfaces, producing inflammation and pustular eruptions of the unabraded skin. The juices of the flesh of an animal that has died of an autochthonous anthrax have produced a fatal erysipelas. In man phagedæna reproduces its like, and will do so repeatedly ; and although it may have supervened upon a syphilitic sore, it may communicate to another person not syphilis, but phagedæna. If this sets in before there

---

* " Evils of Quarantine Laws and Non-existence of Pestilential Contagion." By Charles Maclean, M. D. London, 1825.

has been time for general contamination, it may sometimes prevent constitutional syphilis.*

Again, too much faith should not be placed in any profound and complex reason assigned for the immunity enjoyed after one attack of any disease. Very simple circumstances account for the possible ravages of cystic parasites in the young and tender animal, which are rendered impossible when the tissues acquire a special density by age. Moreover, we might adduce many instances from the inorganic world, and especially from the field of metallurgy, of profound and lasting modifications in the toughness, ductility, resiliency, and so forth, of material, according to the methods of preparation and subsequent treatment.

It is well to remember that complex organic matter is governed by the same laws of energy as those which operate in all other matter. A man stores potential energy in the shape of oxidizable food elements and tissues, and is affecting all around him by the manifestations of a kinetic energy—heat and mechanical work. In health and full vigor he is always expending less matter and energy than he takes in, and provides for growth in the young and almost superabundant deposits in vigorous manhood. These relations can be disturbed in a very simple manner, and such disturbances may result in death or life-long idiosyncrasies.

One of the most simple and most sure methods of instantaneously upsetting the relations between systemic demand and supply, is by modifying the usual normal relations between the tensions of the gases in the blood and in the external atmosphere, respiration consisting in the equalization of those tensions. In virtue of the very low tension of the oxygen of its blood, an animal placed in a confined space can consume almost the whole of the oxygen which that space contains, while the evolution of carbonic acid is very soon stopped by an equalization of the tensions taking place (Wilh. Müller). †

In studying the development of a disease which, whatever may be its specific cause, originates and spreads in foul air, and

---

* "Lectures on Syphilis." Delivered at the Harveian Society, December, 1876. By James R. Lane, F. R. C. S. London, 1878.

† "Elements of Human Physiology." By Dr. L. Hermann. Translated by A. Gamgee, M. D., F. R. S.

the virulence of which apparently bears a direct ratio to this foulness and the confined nature of the space occupied, we have to bear in mind the conditions of normal and abnormal respiration—the conditions of normal and abnormal oxidation throughout the system.

Where and How may Yellow Fever Develop?

An empty ship from the tropical Atlantic, relieved of its crew and cargo, has been closed alongside a wharf in hot weather, and an active decomposition has set in, so that, when ports and hatches were opened, the volume of fetid gases has overwhelmed the bystanders, engendering yellow fever.

More commonly a foul ship entering a harbor from the yellow-fever zone, without manifest sickness among its crew, has been detained under conditions favorable to increasing foulness; and either a ship-keeper, shipping-clerk, or friends of the apparently healthy crew, have gone on board, returned home, and introduced yellow fever in the port. Still more commonly the sickness develops at sea near the equator, when the ship is shut up loaded, and remarkable for numerous *impedimenta*, as well as for being deep in the water. The sailors may be buried at sea, and the ship is placed alongside other foul ships in harbor, or in a slip near some foul dungeons—too bad to be honored by the title of dwellings—and the fetid emanations from the hold are transferred more or less directly into other ships or into these pestilential foci on land. Just according to the facilities of such intercommunication, does the disease become prevalent in the harbor or town, or both. Its inland inroads are practically abortive.

It is, therefore, incontrovertible, and very diverse facts can be adduced in support of the view, that the poison is generated outside the human frame. It originates, as Dr. Sanderson might say, in a quasi-spontaneous manner, in inclosed floating chasms, whence a pronounced rotten-egg fetor is the most obvious characteristic. Never has it been found to originate in vessels confined to inland lakes or rivers. It is always to be found somewhere in the ocean shipping, traversing the Atlantic calm-belts. I shall not prolong this statement by needless repetitions, and my main object at this point is to establish that there are certain undeviating conditions, beyond the sick man on land, under which we can trace and reach the poison, where there can be no pretense that man is carrying it, and only very slight, indeed, for suspecting that man played any part whatever in engendering it.

Now this ship poison is breathed by a susceptible man, and it kills him. It poisons him if the dose be large, perhaps as quickly as a mortal dose of phosphor paste or corrosive sublimate. Like these it produces very definite lesions and symptoms, and like these, I believe, so far as I can determine by all inquiries in my power, it is not reproduced in the system. Some of it may be and possibly is thrown off without change; but if unable to kill the person attacked, it leaves its most marvelous impress on the system, by creating an immunity from subsequent attacks.

The coincidence of cases, as the disease spreads, indicates a pestilential condition of the ship or low-lying dwelling, and not propagation by contagion from the sick to the healthy. Sanitarians, it is true, have been driven to practice isolation and to depend on it mainly, but simply because the presence of the *virulent entity* I have before mentioned, though readily attaching itself to things rather than persons, can only be determined by the sanitarian wherever its physiological effect is obvious. It were well could we trace it before it enters the human system; but all effort so far to ascertain its definite and recognizable existence, apart from the human system, has failed, just as much as the recognition of snake-poison apart from the snake. And probably we may fail throughout all time, as we have failed in recognizing the reproductive animal poisons which flow in with diseased animal secretions or suppurations.

There are the gases of decomposition, which often announce the probable locale of the insidious enemy; but the disease is witnessed spreading where all fetor is lost, or at all events where it is imperceptible, and we have then to trace its operations in the animal economy.

There is organic matter within and without the body, which, with our present means of investigation, seems almost formless protoplasm, capable of motion and endowed with vitality or irritability. Structure in no way indicates that which belongs

to a jelly-fish or a man. Protoplasm is all albuminoid, converting external matter into itself, and being capable of such changes
as will contribute to the firm growth and development of the
higher as well as the lower forms of organic life. This protoplasm is undoubtedly susceptible of lasting as well as transitory
impressions; and what the change is, we know not, which insures
the development of secondaries in syphilis, or the periodic recurrence of ague long after a man has left the marshes.

A formless protoplasm, for aught we know, may be the sea-
ferment in ships growing out of putrid and fermenting bilge-
water, and its action may be almost catalytic—an action by mere
presence, or contact with the blood-protoplasm. This undoubtedly changes, for the blood is affected at once; the arrest of the
higher secretive functions, the instant cessation of molecular
assimilation in the tissues, the retrograde changes to fat in muscles and liver, and the passage of free albumen into the urine,
indicate how profound and unmistakable has been the influence
of that agent which progressively checks and ultimately arrests
all normal nutritive function. So fierce and active may be the
first invasion of a ready victim, that general cessation of function, for that instant of time mentioned by Hirn, seems to be the
method of death. But if the influence is milder and a progressive course be engendered, then the reparative powers surmount
the disintegrating force and conquer.

I may have formed a wrong impression of the possible action
of this poison, but certainly I am not in error that the product
of foul fermentation may elude our perception, and may, without necessarily reproducing itself in the human system, so impress the living albuminoid molecules of the body as to arrest
development and insure molecular and then bodily death.

There is one possible method of action which undoubtedly
operates to some extent in the ship's hold. The active agent is a
deoxidizing compound; and when we consider how carbonic oxide
kills without reproducing itself in the living body, by excluding the possible oxygenation of the blood, we may have a clew
to the operation of this agent.

Dr. Edward C. Seaton, in a preliminary note to the Local
Government Board on Dr. Klein's "Report on Infectious Pneu-
mo-Enteritis of the Pig," said so late as the summer of 1878, that

"a general result of the studies, which have of late years been made by various pathologists, as to the intimate nature of the different zymotic contagia, has been to render it extremely probable that each distinct zymotic disease is, in its essence, an invasion of the animal body by a distinct, extremely minute, living, and self-multiplying thing, biologically specific in the changes of growth which it undergoes, and chemically specific in the fermentations which it can effect; that the infinite multiplication and swarming of this specific organism, in the blood and tissues of the infected body, underlies and causes in each case the more obvious characters—the anatomical and chemical changes, and the so-called symptoms of the disease; and that the infinitely minute germs of the specific organism, barely perceptible with the highest powers of the microscope, constitute, as they pass from animal to animal, the essential means by which the specific disease is communicated and spread."

In yellow fever we have undoubtedly the origin of the disease in putrefaction—hence in a septic ferment, which in all probability only develops in ocean waters under tropical conditions in the Atlantic. In no other sea has the disease been known to infect ships, ports, and harbors. But this in no way indicates a *necessary* dependence on a *living* ferment.

Professor Panum, of Copenhagen, Drs. Bergmann and Schmiedeberg, and Dr. Burdon-Sanderson have proved that a specific chemical body, separable by filtration from any fluid containing it through porous porcelain, not albuminoid in its reactions, and having a remarkable power of generating bacteria in a "cultivation fluid," is the active septic poison which induces the symptoms of a putrid infection. Dr. Sanderson, writing in 1876, states that the facts promulgated by Professor Panum ("Virchow's Archives") are quite irreconcilable with the often too carelessly received assumption, "that the process of septic infection is dependent on the development of a living contagium." The only group of bodies to which this fever-poison may be compared is that of the so-called "unformed ferments."

This suggests the possibility of a very decided and fundamental distinction between the organic and parasitic causes of the pure contagia and the inorganic "sepsin" or "septine,"

10

which may play a conspicuous part in the development of such diseases as typhus and yellow fever.

We must not allow these investigations into ultimate causes to bewilder or disturb us. We can stamp out the disease without the microscope or profound chemical research, though an intimate acquaintance with the pyrogenic agency might simplify our work. Whether the body be organic or inorganic, bacteria may have, and most probably do have, a very important purpose to serve as carriers of the agent—starters of decomposition; and heat is essential to their motion, to the activity of their production, and to the ready transmission of the disease. Chilled and motionless, they are inert; and once the yellow-fever poison sinks to the soil, or is subjected to a temperature below 32°, it is gone at once and for always. This is the one great and positive fact which the history of yellow fever has invariably revealed.

The known action both of heat and cold on the yellow-fever poison would indicate that bacteria can not be the ultimate essential elements in the production of yellow fever; for while permanent spores resist temperatures up to 212° Fahr. (Schwann), 230° Fahr. (Pasteur), and even 266° according to Schrader, bacteria were reduced down to 0° Fahr. by Cohn without killing them, though their movements and reproductive powers or activity as ferments ceased till again heated. By the aid of liquefied carbonic acid, Frisch has lowered the temperature of bacteria down to −124·6° Fahr., but on warming them again he has developed coccus and further bacteria. Schumacher has, however, said that sudden and extreme lowering of temperature changes the organisms. The degree of heat for greatest activity of development is 95° Fahr. according to Onimus. The bacteria were supposed to be reproduced both by spores and subdivision. The spores or permanent germs capable of resisting the high temperatures above stated have been regarded as constituting the principal sources for the dissemination of these inferior organisms.* To bacteria we owe the prompt reduction of dead organic to inorganic matter.

So far had I written when I had an opportunity of reading Dr. J. J. Woodward's "Address on Dysentery and Bacteria,"

* "Les Bactéries," Thèse par le Dr. Ant. Magnin, Paris, 1878.

published in 1878, and a profoundly instructive paper by Dr. Timothy Richard Lewis, on "The Microscophytes which have been found in the Blood, and their Relation to Disease." Dr. Woodward has failed to find "the relation of bacteria, and especially of spherical bacteria, to dysentery." He asks, "Is it at all reasonable to believe that these vegetable rods or granules, always present in such countless numbers in the healthy stools, are the occult cause of the diphtheritic process in the intestine? Certainly, I must believe that if the case of dysentery were the only case of the kind, all men knowing the facts would answer in the negative."

There is no evidence with regard to yellow fever that bacteria of any kind produce it. The effect of temperature on spores and bacteria indicates that they can not, as it has been supposed, be operating as different conditions of the same organism in the production of yellow fever, which is arrested both by steam-heat and ice-cold.

The abundant development of bacteria demands oxygen, but this is reduced to a minimum, or is absent, where the most virulent form of yellow-fever development may be advancing in the holds of ships. Wherever nature is favoring that cycle of operations, between the organic and inorganic world, so essential to human life, whether in the West Indian marsh-lands or African jungle, prompt and abundant oxidation, and the undoubtedly unfettered development of untold myriads of bacteria, are of necessity concurrent phenomena. There no yellow-fever poison forms nor has its origin. This is my conviction, after a most careful search for evidence ; and if the conclusion may be deemed by some somewhat far-reached, it is impossible to deny that yellow fever and bacteria are best developed, in the *first* case, where dangerous deoxidation has rendered air irrespirable, and in the *second*, where the fresh breeze of sea or abundant air contributes to an active vitality.

Dr. Lewis in his admirable and logical exposition shows, according to Nägeli, that the forms of plant-life, which have been mostly recognized as having been more or less closely associated with changes in living animal substances, are the lower kinds of fungi.

Nägeli separates these into three groups :

1. *Molds*—characterized by branched, segmented, or unsegmented filaments.

2. *Sprouting fungi*—yeast-cells of various kinds, consisting of more or less oval corpuscles, which multiply by means of sprouts from their surfaces.

3. *Cleft fungi* or schizomycetes—minute spherical or oval bodies, which are multiplied by fission only, and which sometimes remain isolated, at others form unbranched rows (rods, threads, etc.), but only occasionally present a cubiform aspect. To this group the bacterium, vibrio, vibrio-bacillus, spirillum, etc., belong.

Germs Unrecognizable.

Virchow,* a strong advocate of the germ theory of disease, in attempting to explain away the apparent identity of vegetable forms in healthy and dysenteric stools, says: "Although we may be unable to see these inner differences in such minute bodies as vibrios and bacteria, yet we ought to remember that in case of the formative cells of the ovule, and of numerous pathological growths, although they appear as gigantic forms alongside of vibrios, we are unable to predict what will grow out of them. Yea, eggs themselves are often so similar to each other, that the differences of the animals which will proceed from them can not be suspected in the least degree."

Nägeli has pointed out the serious and almost unavoidable fallacies attending conclusions from cultivation experiments. The assertions, such as Schumacher's, as to the transmutations of organisms, are unsupported by adequate evidence. A fungus, according to De Bary, undergoes but a very limited and well-defined range of changes. The mold and sprout fungi are closely related, but with one exception—they have not yet been seen to pass from one form into the other (Nägeli). Fission fungi do *not* germinate, so that the spores which resist high heat and the bacteria which live at the lowest possible temperatures are independent organisms.

Bacteria, vibriones, and bacilli are commonly detected in the muco-salivary fluid from the mouths of healthy persons. Dr. Beale, writing in 1870, said: " The higher life is, I think, every-

* See Dr. Woodward's Address before the University of Pennsylvania in 1878.

where interpenetrated, as it were, by the lowest life." He considered there was no evidence that "bacteria are really the active agents, in cases in which the blood has been shown to exhibit the properties of a specific contagious virus"; and he asserted that "it is difficult to imagine anything further removed from the fact than the statement that the dust of our air consists of disease germs."

Dr. Beale's judgment may not have been infallible, but he has rendered excellent service, and uttered some wholesome cautions, in opposition to the boldest speculators on the question.

Lower Organisms not specifically injurious, nor structurally characteristic of Fevers.

Pasteur—the greatest mind grappling with the difficulties of living ferment investigations—has stated that the blood in health is absolutely free from the microphytes; and Drs. Douglas Cunningham, and Lewis were able to satisfy themselves some years ago that bacteria, vibriones, bacilli, and so forth, very speedily disappear from the liquor sanguinis, even when introduced into it during life in considerable numbers. Their presence in appreciable numbers is, judging from experience, incompatible with a state of perfect health. The same remark does not hold good as regards parasites of, apparently, animal nature. In certain diseased conditions microphytes are very generally present—not invariably, nor proportionate to the gravity of the malady.

Dr. Lewis refers to the fact that any organic substance, added to suitable urine, would be followed by a crop of bacillus. He once tested this and found it to be the case. "It need hardly be added that organisms thus obtained would produce no effect on animals if freed from the decomposed urine."

Dr. Lewis states most forcibly the grounds for forbidding the adoption of the doctrine of a germ theory of disease, indicating (1) that these organisms, as ordinarily met with, are not specifically injurious when introduced into the animal economy; or (2) that the forms found in disease are not proved to be morphologically different from those known to be innocuous.[*]

[*] "Quarterly Journal of the Microscopical Society," July, 1879.

The attractive theory of *germs* or *spores*, inducing yellow fever, is leading physicians and journalists to develop some decidedly false theories and views of the disease; so that it is high time to point out that, without attempting to solve the questions arising from discussions of the "vital" and "physico-chemical" theories of fermentation, we know that yellow fever originates in putrefaction—is nourished by decay and dirt; and that this peculiar decay or putrefaction is arrested and killed by almost any temperature below 32° Fahr., and probably below still higher degrees of heat. Dry, cold air is its great enemy, and that is what the engineer can provide in every ship that floats, and every home invaded by the pestilence.

CHAPTER V.

In the summer of 1878 I was in St. Louis, and, with the outbreak of yellow fever in the Mississippi Valley, it was perfectly clear that general medical knowledge and the most heroic zeal stood but poorly in stead of a definite and comprehensive understanding of the origin and nature of the disease. I heard enough of the evils and brutalities of inland quarantines, though these were mild and innocent compared with the tortures endured in the lazarettos which Chervin abolished. I can not resist quoting a paragraph from a St. Louis newspaper, which I cut out last summer, but without name and date :

Making Quarantines Pay.

An excuse was sought for, about a week ago, by the authorities of a number of cities and towns not far distant, to quarantine against St. Louis on account of yellow fever—although it was known by all intelligent people in those cities and towns, as well as in St. Louis, that no genuine case of yellow fever ever existed in this city, and the plague did not exist here at all, as an epidemic or otherwise. In view of which, the question very naturally arises, whether the authorities of the cities and towns in question had not, aside from protecting themselves, an eye out for the "main chance." The "New Orleans Times" has an illustration in point:

"Take the case of Washington, in St. Landry parish, for example—only for example ; there are a dozen or two more of the same sort. Washington quarantines against New Orleans, prohibiting passengers, freight, and everything. The local newspaper, with a sweet and childlike trust, too heavenly for this cold world, expresses the hope that 'our merchants will be fair and considerate to the people' while the embargo lasts ; but it turns out that the presiding official is a leading grocer, and that others of the authorities are dealers in articles which a quarantine will enhance, and the next thing we hear from Washington is that coffee has gone up to forty cents a pound. Of course these towns have a deeper interest in a tight and binding quarantine than New Orleans can possibly have. It enables the prominent merchants to work off their old stocks at famine prices, and to make a general clearing up on terms of bewildering advantage to themselves. We do not expect them to listen to our arguments in the matter of merchandise. It pays to be mortally frightened under such circumstances, and we know human nature too well to waste time in expostulation."

I then felt what I now know, that to permit such a disease to penetrate so far up that splendid river was a crime originating, as all crimes must, in ignorance and cupidity.

I reached Washington in September, and frequent interviews after that with Surgeon-General Woodworth convinced me that the non-intercourse system, in a trading community, was permanently impossible, but that much might be done to limit quarantine if ships could be disinfected. The only agent known to be the real destroyer of yellow-fever poison was *cold*, and my practical familiarity with the subject of artificial refrigeration led me to pen the following statement, at Dr. Woodworth's request, on the 23d of last December, which he read before a meeting of the Board of Experts held in New Orleans :

RIGGS HOUSE, WASHINGTON, D. C., *December* 23, 1878.

Destruction of Yellow-Fever Germs by Cold.

The disinfection of delicate and perishable merchandise, the demand for some efficacious means of purifying a ship some distance from land, and the difficulties incidental to the purification of the bilge, have suggested the construction of a steamboat or barge for these and similar objects.

The approach of a suspicious craft, hailing from infected ports, might suggest precautionary measures, notwithstanding the health of crews, proof against yellow fever and the pernicious influences of malaria. The brief delay and expense of lowering the temperature of everything on board, during the dangerous season, would be regarded as of little moment compared to what might be deemed unfair exclusion from markets. Ships laden with fruit, however delicate, would not suffer if the whole were brought down to 32° F., and at this temperature the contagion of yellow fever is killed.

It is therefore proposed to design and construct a boat, requiring a minimum space and fuel for its engines, capable of steaming out into the Gulf of Mexico or up the Mississippi. This ship will be provided with every convenience for a medical or sanitary inspector, and the necessary help to carry out the process of purification by cold. An ice-machine, capable of lowering the temperature of an entire ship down to 10° or 20° below freezing-point, if necessary, will be erected in the disinfecting steamer, and its practical application will be as follows:

1. The blowing of cold air at any desired temperature into every part of the ship. Cold air gravitates and permeates almost like water, and, by suitable flexible appliances, streams of intensely cold air will be directed to the lower and most inaccessible parts. This air, heated, will ascend, and, being aspirated by suitable fans or blowers, will be purified and passed round and round, till the desired purity and temperature are obtained.

It is obvious that microscopical and chemical tests can be applied to this air in circuit, at first for the purposes of scientific observation, but afterward, probably with the best effect, as a means of determining the influence of the refrigeration.

I desire to direct special attention to Dr. Angus Smith's paper on Ammonia in the "Chemical News" of July 26, 1878, and to the ready method there described for applying the Nessler test to the atmosphere, which may be combined with the well-known glycerine slides for microscopical examination with a view to the detection of organic matter.

2. Pumps will be provided to clean out the bilge, and then wash this out by antiseptic salts, of which perhaps none would be better than chloride of magnesium, which may be obtained in any quantity from sea-water by freezing out other salts and water, and is itself uncongealable. The use of concentrated sea-water at say 15° F. for this purpose, would preclude the use of costly and unpleasant disinfectants, which by their odor and fumes might damage cargo and injure the ship.

3. In the case of fruit-bearing schooners and other ships, the purifying cold process would admirably prepare the cargo for storage on land. In a cold fruit store any quantity of imported produce might be kept harmless and wholesome.

Attempts have been made with ice to cool a ship in the port of New York; but when the work to be done for effectual refrigeration of the hold of a large vessel is considered, it is obviously futile to attempt any such means.* Any ice-apparatus, calculated to lower the temperature of a ship, would have to be used with salt, and by no known means could the quantity of ice and salt required to overcome the heat of a ship absorb that heat rapidly enough to obtain a satisfactory and economical result.

No substantial truth in relation to disease-prevention has ever been more faithfully or justly accepted than that the exposure of naval vessels to the intense cold of northern winters purified them. Apparent exceptions to this have been always recognized, but such exceptions rather tend to prove the rule;

* Dr. A. N. Bell informs me that in the U. S. S. Susquehanna, on which it is supposed a fair experiment in refrigeration for the destruction of yellow-fever poison was conducted, only *nine* tons of ice were used, at a cost to the Government of twenty thousand dollars!

and a confirmed yellow-fever ship has been recognized as one entirely devoid of means for adequate ventilation and permanent sanitation on a long cruise in the tropics. The history and description of those ships which have resisted any disinfectants, and in which disease reappears notwithstanding the application of processes adequate for other ships, must be investigated and written in future for the guidance of the sanitarian. Such ships will be found peculiarly favorable to stagnation of air, old and decayed, and liable in the tropical belt to the development of yellow fever under ordinary systems of ventilation.

Dr. Faget quotes Dr. Carpenter, who, in his sketches from the history of yellow fever in New Orleans, expresses the remarkable opinion, which at the time attracted but little attention (*presque inaperçue*), that " *as far as we know, low temperature is the only agency that can be relied on safely to destroy the infection of this disease.*"

" Let us hope," says Dr. Faget, "that the future will benefit by this remark." " For my part," he says, "I am persuaded that in New Orleans, *when ice has formed in winter, a new importation by the shipping* [the italics are his] is indispensable for a new epidemic; and that, on the contrary, the *absence of cold* in the winter following an epidemic may permit, the following year, a new eruption of yellow fever without fresh importation. For instance: In 1854 we had a small epidemic, which was very severe elsewhere. Not knowing whether there was any frost in the winter of 1853–'54, the yellow fever of the latter year might occur without importation as a heritage of 1853. I am simply citing a supposititious case, for it might by no means be impossible to obtain proof of an importation in 1854 as well as 1853."

It is thus that matters occur in Vera Cruz, Havana, the Antilles, where indigenous ice is unknown, and where the temperature of the air is never lowered to the freezing-point; in such localities an indefinite series of epidemics is possible after a single importation.

Artificial Refrigeration proposed by Dr. J. C. Faget.

Tersely, suggestively, and with no doubtful expression, Dr. Faget penned the following in 1864, soon after the importation into New Orleans of the Carré ammonia machines, which have continued in operation ever since:

"It is positive, and tradition has always recognized the fact, that the cessation of our epidemics of yellow fever is with the first white frost.

"This fact, that frost puts an end suddenly (*tout à coup*) to our epidemics, naturally leads us to think that the germ of these epidemics is therefore destroyed by the lowering of temperature, provided it reaches to 0° C. Now, *art* never can do better than to imitate nature.

"The lowering of the temperature in the holds of ships which have to be purified should be tried. Might not frigorific mixtures be obtained at low cost? To-day, that they have succeeded in manufacturing ice on a large scale and at little cost, *even in New Orleans*, in midsummer, can't we foresee there a means to destroy the germ of yellow fever in the holds of ships which import this terrible plague? I give this idea for what it is worth ; it seems to me worthy of attention."

Cold as an Antidote endorsed by the Board of Experts.

This excellent instruction to sanitary reformers fell flat; but the Board of Experts recently uttered no doubtful echo to Dr. Faget's views and sentiments when they said : "Yellow fever, in its epidemic form, is a disease of warm climates, and of warm seasons of the year. The yellow-fever poison is not able to withstand the influence of frost, and, when exposed to a freezing temperature, it is rendered innocuous, and is probably destroyed."

Having had an opportunity of explaining my system of refrigeration and disinfection to the Board of Experts, special reference is made to this at page 27 of the published conclusions. They say : "Research and experiment in reference to disinfection should be liberally encouraged. This is particularly true in respect to those experiments which aim at the practical destruction of yellow-fever and cholera poison, by artificially produced extremes of temperature, either of cold or heat. If the apparatus and experiments now projected for the utilization of extreme cold to this purpose should be found to be of practical application to the disinfection of the holds and other parts of vessels, their success would prove to be a sanitary acquisition of inestimable value. Should this method of disinfection prove impracticable in its application to holds of vessels, it may still prove more efficient in destroying the infection of clothing, baggage, and many kinds of goods, than any other artificial means which can be employed."

It is not my intention at present to give a description of the elaborate designs which have been developed since my first conference with Dr. Woodworth, but the opinion of the Board of Experts encouraged me to spare no pains and no expense to obtain the most perfect possible result ; and the National Board of Health has, since Dr. Woodworth's death, endorsed my measures and honored me with confidence.

Here I desire to pay a just tribute of recognition to Senator Isham G. Harris, Chairman of the Senate Committee on Epidemic Diseases. Startled at first by the announcement of the bare possibility that we could refrigerate infected ships in the Gulf of Mexico, he spared no pains and lost no opportunity to satisfy himself of the real merits of the project, and of my fitness to conduct the experiments. He searched till he felt sufficiently convinced to endorse my plan. From first to last he was steadfast, and so loyal to the cause that, if the experiments, demanding as they did a not unimportant appropriation, are carried out, and result, as they are sure to, in success, it should never be forgotten that to Senator Harris is due the passage of the following short act, the first measure approved by the President during the extra session of 1879 :

Be it enacted, etc., That the Secretary of the Treasury be, and hereby is, authorized to contract for the purchase or construction of such steam-vessel and refrigerating machinery, or to arrange with the Navy Department for the use of such vessel, as may be recommended by the National Board of Health, to disinfect vessels and cargoes from ports suspected of infection with yellow fever or other contagious diseases, the construction of the same, if such construction shall be recommended by said Board of Health, to be under the inspection of an officer of the Bureau of Steam Engineering of the Navy,who may, at the request of the Secretary of the Treasury, be detailed by the Secretary of the Navy for that purpose ; and for the purpose of such purchase or construction the sum of $200,000, or so much thereof as may be necessary, to be immediately available, is hereby appropriated out of any moneys in the Treasury not otherwise appropriated.

Approved April 18, 1879.

When the measure for the construction of my refrigerating ship was before the Senate, Governor Harris was fortified by the following expression of opinion:

WASHINGTON, D. C., *March* 27, 1879.

SIR: In reply to your letter of the 26th instant, requesting that we should give you our opinion as to the merits of Professor John Gamgee's proposed refrigerating steam-vessel as a means of disinfecting vessels and cargoes, we have the honor to report as follows:

1st. We have examined the models and the drawings relating to this apparatus; have listened to the explanations given by Professor Gamgee and the engineers who have been engaged with him upon this subject.

2d. We are of opinion that the apparatus, if properly constructed and managed, will produce a temperature below the freezing point of water, probably as low as zero Fahrenheit.

3d. In an empty ship it is probable that this low temperature can be produced throughout by the refrigerating vessel, if sufficient time be allowed for the working of the apparatus.

4th. In a ship with a cargo in it, it will be more difficult to produce this uniform low temperature; and it is probable that in almost all cases it will be necessary to discharge the cargo before a ship can be properly disinfected by this or any other process.

5th. Whether the production of a temperature, even as low as zero, in a ship, for a comparatively short period of time, *id est*, for a few hours, will destroy or render permanently harmless the yellow-fever poison, we do not know, there being no satisfactory evidence in existence upon this point. But from what is known as to the effects of temperature upon this disease, we are of the opinion that the experiment is worth trying.

6th. It will be seen that we consider the project as an experiment; it is in fact impossible to view it otherwise.

But even in the case of failure the results can hardly fail to be of considerable scientific value, while in case of success the cost would be a comparative trifle.

We are respectfully, etc.,

THOMAS J. TURNER,
Medical Inspector United States Navy, member of National Board of Health.

J. B. HAMILTON,
Surgeon United States Marine Hospital Service, Treasury Department.

J. B. BILLINGS,
Surgeon United States Army.

Hon. ISHAM G. HARRIS,
Senator and Chairman of the Committee on Epidemic Diseases.

The discussion in the House indicated the faith reposed by Southern members in the effects of frost as a preventive. General King, of Louisiana, said:

Gentlemen who have witnessed the ravages of yellow fever can only appreciate the necessity which is felt in that portion of the country of keeping out an enemy so subtle and so destructive. We know that during the hot weather that disease spreads and carries devastation with it on all hands. We know that on the first day of frost a shout of relief goes up, quarantine is broken, and commerce once more begins to flow in its usual channel; and we know also that, with the announcement in springtime of the approach of the yellow fever, terror is at once depicted in the countenances of the inhabitants of that part of the country. On the application of the Board of Health of the city of New Orleans, I therefore urge this House to take into earnest consideration measures to prevent the introduction of this enemy into our midst again. They approve of this freezing process. It may be suggested that this is experimental; but it is in the right direction. The question is, how far will this Government exert its power for the protection of its citizens against this terrible disease? Millions are spent in the destruction of men. Is it not time that, by the authority of a civilized people, thus represented, thousands at least should be spent for the protection and preservation of human life and happiness?

The Hon. John Goode, of Virginia, followed with much earnestness, and remarked, *inter alia:*

Now it has been said here that this proposition is endorsed by the National Board of Health—a board selected by the President, a board composed of the most eminent sanitarians in this country; and not only endorsed by the Board of Health, but it has the approval of the most eminent scientific men in the land. The whole bill proceeds upon the basis that the yellow fever is an exotic brought to this country from tropical climates in the holds of vessels, and the germs from which it is propagated may be effectually frozen out and destroyed by the refrigerating process. In other words, it is now a generally conceded fact that the yellow fever can not live in a freezing temperature; and, as my friend from Louisiana [Mr. King] has just remarked, so well is this fact known in the South that the good-people in the churches always pray for frost, and when frost appears there is a shout of joy throughout the fever-stricken localities.

General C. E. Hooker, of Mississippi, said:

We propose to try this experiment with nature's great teachings as our example. We are, in other words, endeavoring to do for every vessel that comes into our ports from an infected region that which nature does where her laws do not interfere with any of the comforts, conveniences, and necessities of man; and I say that experience has demonstrated that wherever a certain amount of cold is reached, there we believe the disease, though decimating by thousands, will be disposed of, and general health prevail in the cities.

The Hon. J. G. Updegraff, of Ohio, added :

In some sense any plan of quarantine, and especially any plan to keep out of the country a terrible epidemic, is an experiment. But these basis-facts are established : first, that yellow fever is not a native of this coun-try, but is an exotic brought to us from tropic climates under certain con-ditions, one of which is heat; second, it is well established that cold is the destroyer, the antidote, of yellow fever. This has been well settled since this disease prevailed first in America. Nearly one hundred years ago Dr. Rush connected its prevalence with hot weather and its disappearance with cold.

This proposition is to use a refrigerating ship to freeze out and purify vessels coming into our ports, as recommended by the National Board of Health.

Dr. George B. Loring, of Massachusetts, in an eloquent ap-peal, stated :

I wish to bear my testimony to the value of all such endeavors and ex-periments to suppress contagious and infectious disease in this land. There are certain facts, Mr. Speaker, known with regard to the control of yellow fever, about which there can be no doubt, and one is that frost and cold will kill it. The experiment tried on board the Plymouth furnishes no exception to the rule, and the accidents which befell it then cast no reflec-tions upon it. I consider it, therefore, fixed and established that diseases of this sort can be controlled by proper application of cold. So much for acknowledged facts.

Now, sir, in this age of experiment, moreover, is it not the duty of the Government to institute all proper experiments for the control and suppres-sion of ravages of this description? Why, it is manifestly so. The whole history of science shows us that it is the duty of the Government to en-courage scientific investigations for the suppression of destructive contagious and infectious diseases. In all its operations science has gone on, in our day, proving that man has in his hand the power to control those diseases in the animal and vegetable kingdom which are injurious to the community and destructive to life.

How is it proposed to deal with Shipping?

My observations on this point shall be preceded by some very practical remarks by Dr. Vanderpoel on quarantine. He says that " all we know of the yellow-fever germ teaches us that it propagates with the greatest rapidity whenever the ele-ments of heat, moisture, fermentation, and no circulation of air are present. These elements are eminently combined in ves-sels coming from the tropics. Their cargo is sugar or melado ;

they are put on board under a tropical sun ; both are dirty cargoes, and more or less leakage goes on into the bilge ; the
hatches, too, are tightly battened. If the germs are on the vessel, can there be conditions more favorable than these for their
active increase ? Experience daily teaches us that on such vessels, should the voyage be prolonged, cases of sickness follow
each other in rapid succession, until the vessel becomes a very
pest-house ; whereas, if the voyage is completed in eight or ten
days, we find the sickness breaking out but a day or two before
entering port, where prompt measures put a stop to its further
progress. The longer such a vessel is allowed to remain un-
touched, i. e., the longer a quarantine is exacted, the greater
will be the virulence of the disease, and the greater the danger
of spreading the infection. The epidemic which appeared at
Bay Ridge, Long Island, some years since, was undoubtedly due
to this cause. Vessels were strung along the Long Island shore
to ride out a specific quarantine, until the whole fleet was thoroughly infected, and the disease passed readily to the main-
land. It is this practice of detention which in the past has
contravened the interests of shipping and commercial men, and
has kept up a chronic warfare between merchants and quarantine." *

No words could better unfold the dangers of commercial ob-
structions in the nature of protracted quarantine, nor suggest
more cogently the radical methods to be adopted for instant
and absolute purification.

Dr. Faget, writing in 1859, said: " Quarantine measures
must aim at destroying the morbid principle—the germs of
the disease. Now this is not a question of time; it is a question of instant action by all the means of purification, and this
after the complete unloading of the ship." These means of
purification he defines, in September, 1859, as currents of air
and water, submersion, refrigeration, chlorine fumes. " Merchandise," he says, "can be very easily disinfected. Passengers can be treated very rapidly, provided their goods are thoroughly purified."

* " Quarantine," by S. Oakley Vanderpoel, M. D., LL. D., New York, p. 20.

Little Danger from Cargo.

Dr. Vanderpoel states that "the experience of many years at this port—and in this I am confirmed by the observations of the St. Nazaire epidemic in 1862, and others in Spain and Portugal—has led to the conclusion that little danger of transmission arises from the cargo proper. The clothing and effects of the individual, and the dark recesses of the hold, are the favorite lurking-places of the poison."

The doctor briefly describes his practice as follows (p. 21) :

"In the case of the city of New York, the vessel is sent to the upper bay for the purpose of discharging the cargo, the spot being nearly two miles distant from the nearest shore, where the bay is five or six miles wide. The crew of the vessel is usually discharged before the cargo is broken. Stevedores, coopers, and all who work in the hold at this discharge, reside in hulks anchored near by, and are not allowed to return to the city until a period of five days has elapsed since they worked on a suspected vessel. The cargo is swung upon open lighters, and then carried to storehouses about three miles distant. No special precaution is exacted of the lightermen, except that they are not allowed to go on the vessel which is unladen ; nor is the cargo subjected to any restrictions concerning the warehousing. During the discharge of the cargo, fumigations with chlorine should be made once or twice daily.

"The vessel once emptied of her cargo, the process of purification begins. With steamers this is comparatively speedy and easy of accomplishment. A fire-hose is attached to the force-pump, carried to the hold, and a full head of water thrown into every part of it. At the same time men are set to work with scrub-brooms, who work until every portion has been scrubbed and is as clean as a housewife's kitchen. A discharge-pump is kept working at the same time, and the water is poured in, not only until thorough scrubbing is completed, but until the discharge-water is as clean as that which enters. Hatches are left off, ports are opened, and the fullest airing is given until the vessel is dry ; then everything is closed, and fumigation is started in every part of the vessel.

"Pratique is then given, and the vessel returns to commerce.

11

" With sailing vessels the process of purification is slower, on account of the absence of the steam-pump. It is often found advisable to have a tug-boat lie alongside, so that use may be made of its force-pump and hose in the same manner as on a steamer. The subsequent measures to be carried out are the same in both kinds of vessels. In other cases dependence must be placed upon the ordinary hand deck-pump; the purification and cleansing, though occupying a longer time, can be made equally effective. I place far more reliance upon the liberal use of water, the scrubbing until the vessel is absolutely clean, the pouring out of clean water from the bilge, than upon any process of mere disinfection and fumigation. These measures, which are simple and easy of application, are, I believe, when thoroughly carried out, perfectly efficient."

Dr. Vanderpoel is only half right in a sentence which was adduced in Congress in opposition to the construction of my refrigerating ship. He says (p. 22): " While all such work must be thorough, we should not lose sight of the economical feature in the process. A method of purification, which may be unobjectionable in principle, may also·be expensive, and so complicated in application as to cause a tax which no commerce could sustain, and which would practically work its entire obstruction. So, while I recognize the efficiency of steam, great heat, or intense cold, as agents which all attain *with certainty* the desired result, I recognize also that their practical application is enormously expensive, requiring complicated machinery and a corps of skilled men, and does not insure results that are better or more certain than those attained by the more simple processes."

Note Dr. Vanderpoel's declaration that with sailing vessels " *it is often found advisable to have a tug-boat lie alongside.*" But if a tug is wanted for washing, is it not most desirable to clear the hold, at the very beginning, of the pestilential air, which has so often overcome the workmen engaged in purifying a ship? Moreover, in warmer climates the tendency to infect the tugs, by the foul air of the infected vessel, is a much more serious cause of alarm than in the port of New York, where the atmospheric conditions are naturally inimical to the ready propagation of the disorder.

The cost of thoroughly disinfecting ships in New York will be far less, moreover, than the present cost of inspection and necessary detention. On the score of economy everywhere, if we must save the pence to take care of the pounds, a skilled and competent staff of disinfectors on board a proper disinfecting ship, or floating laboratory, must be recognized as preferable to the half measures which have proved so illusory.

Thorough Disinfection Essential.

The epidemic of 1879, especially in New York, demonstrates that it is not safe to rely on partial disinfection by chemical agents, and giving ready pratique from infected ships, if we wish to avoid cases of sickness, though isolated, and a (perhaps senseless) panic. The disease pierces with great difficulty through such barriers as Dr. Vanderpoel has at his command; but he has felt compelled, by an unusual gathering of foul shipping, to allay public fears by more stringent measures, and measures which should hereafter be rendered needless by a wholesome supervision of vessels in foreign ports, and effectual disinfection off land.

As I am writing, August 7, the " New York Herald " reaches me, and contains the annexed paragraph :

The owners of steamers engaged in the Havana trade have concluded to suspend the regular trips of their vessels, and unite in sending their freights by one or another of the regular lines alternately, pending the present quarantine regulations, which forbid them to discharge their vessels or take in cargoes at the docks. The steamship Niagara, now at Upper Quarantine, will probably be the first vessel loaded, and leave for Havana upon the new plan.

Dr. Vanderpoel has issued orders to the officer in charge of Lower Quarantine to permit no sailing vessels from infected ports to come up until further instructions. This order has been issued in consequence of the accumulation of a large fleet of vessels at the discharging grounds, between Robbin's Reef Light and Bedloe's Island. There are as many there now as the stevedores can manage, and until the fleet is thinned out no more can be admitted. The steamships are not included in this order. They will be allowed to come up, after undergoing the usual quarantine regulations, as heretofore.

Two other items from the same paper indicate how much a rational national or international organization everywhere is needed. They read as follows :

Only five feet of water in the hold of the hospital ship Illinois, and eleven souls on board. This is the brief comment on the leaky hulk that stands guard at our Lower Quarantine. Truly, what a careful set of health officials we have in New York!

THE HOSPITAL SHIP SINKING.—The Illinois, which is anchored in the lower bay and was originally intended for the reception of invalids from infected vessels, but is now used as a boarding station from which all vessels from infected ports are visited, was yesterday in a sinking condition. The vessel is a very old one; contact with ice has stripped her of all her sheathing, and her seams have opened as she has gradually become worm-eaten. For several days past she has been leaking, and yesterday there was five feet of water in her hold. On board were Dr. McCartney and eleven other persons, including women and children. Quarantine Commissioners McQuade and Oakley visited the Illinois in the afternoon, and, in answer to a dispatch, a large schooner of the Coast Wrecking Company, in tow of a tug, started for Lower Quarantine. It is thought that when the Illinois is once pumped out her own pumps will be equal to the task of keeping her free. She has not been in a dry dock for eight years.

Why not Attack the Disease off Land?

The first practical question, which started the idea of the refrigerating ship, came from a New Orleans steamship owner, who wished to know how to disinfect the bilge, and otherwise purify a ship, without pulling her to pieces or destroying her fittings, etc., by steam at high temperature. Information soon reached me that a yellow-fever ship often approached port flying signals of distress, and communication with land was forthwith established by a pilot and tug-boat—both unable to do more than take the vessel to quarantine, where the pilot and tug-boat crew would not care to remain, and might therefore carry in their clothes or vessel the infection to the city from the stricken ship.

The *non-intercourse system* I have advocated consists in having special ships, special officers, and special crews trained, as in the case of fire-ships, for special service. Can anything be more judicious or more reasonable? I have striven to explain this question to all inquirers, and I am glad that the course of my studies has led me to a more thorough discussion of the whole question from a general scientific and practical standpoint, so that the *raison d'être* of my refrigerating-ship experiments may be the better appreciated.

A most ignorant question has been urged against the construction of my refrigerating ship, viz. : What can one ship do against all the ships on the Atlantic coast ? If we succeed with this first experiment, will the people hesitate to protect themselves on the ground of cost ? Will any seaport be satisfied to remain without the means of preventing yellow fever ? *To succeed, you would require a fleet of refrigerating ships*, I am told; and I admit that the safety of human life on the Atlantic seaboard of America demands, for some years at least, and perhaps permanently, a floating sanitary police attached to the principal ports, capable of enforcing sanitary regulations and succoring the afflicted. This can only be subservient to and in aid of the quarantine officers on land; but even in the port of New York, if a ship comes in from Havana or Rio, what harm can come of an hour's detention to connect a double hose, of adequate capacity, and blow through the ship fifty or a hundred thousand cubic feet of cool air per minute, and straightway wash out the bilge with pure and cooled sea-water ? Any sick may be taken at once to quarantine, and the healthy with their clothing and baggage subjected to such purification as may warrant their landing. The cargo can then be landed, and the best place for it is in good, economical, cold stores, where it can be subjected to a definite blast of pure air at any desired temperature and for any required time. The ship then can be cleaned, ventilated, and literally frozen out, according to its size, for an expense varying from fifty to one hundred dollars ; and it is quite certain that such purification and refrigeration as I have proposed would leave nothing behind.

The Classification of Ships.

The facts revealed in the foregoing pages prove, as the case of the Plymouth recently proved, that it may be very dangerous to reload an old ship with its decayed timbers, and send her sailing down toward the equator, with a susceptible human freight, and especially should a storm overtake her, compelling the fastening of the ports and hatches.

But, in the first place, international regulations should be adopted similar to those relating to load-lines and seaworthiness. No ship should be allowed to trade in the tropics without adequate ventilating appliances for all weathers. There is no practical difficulty in this, and the great benefits derived from the trifling expense incurred in providing and fitting such appliances would handsomely repay all, and especially those engaged in carrying perishable products. Every steamer could easily provide for the air that is to be drawn through its hold being cooled and even dried.

So essential is it that no unreasonable demand be made on the mercantile marine, that measures should be matured and improved by all the governments interested, so as to command willing and even eager acquiescence on the part of ship-owners.

All ships that are not worth ventilating should be condemned and ordered off the lines of trade. This would be no hardship. It would soon result in a new and fine class of vessels, classed A 1 under a sanitary code, and which would cause little trouble in the future.

Special Ships for Special Cargo.

People in the tropics are not disposed to change from their ordinary practices. They run little risk compared to the people of healthy seaports, and no improvement is made in the methods of transporting such produce as is calculated to foul a ship. But the time has arrived for native northern ingenuity to design ships for the West India trade which will prevent the ready leakage of fermentable matter into the bilge. Iron tanks and wells may in some cases serve, and in every case the ventilation of the bilge must and can be secured.

The paltry objections, constantly made against condemning a lot of crazy, plague-engendering vessels, are not worthy of consideration when the millions annually lost by interrupted and actually mortified commerce are considered. The seaports first, the ships next. No town can protect itself against yellow fever but by excluding it! Let this be remembered hereafter. What can be the meaning of the subjoined extract from a report of the Board of Sanitary Commissioners of Savannah, Georgia, on certain statements in reference to the sanitary condition of that city by Sanitary Inspector A. N. Bell, M. D. ?—" The assumption of Dr. Bell that 'the people do not realize their danger, and do not believe that they can have yellow fever unless it is brought to them from somewhere else, and in the seaport towns they insist that the disease never reaches them except by ship,' is without foundation, and is a misrepresentation of the views and opinions entertained by this board, as well as the opinions of a large number of practicing physicians in this city."

The Board of Sanitary Commissioners were quite justified in believing (if they so believed) that yellow fever would not enter seaport towns except by ship. This is the rule, and at all events, indirectly if not directly, it can only reach them by ship. Dr. Bell has written so well on this question in the past, that I must confess myself utterly puzzled by this extraordinary paragraph.

The Board were quite safe in thundering forth the following challenge. They say: " We challenge Dr. Bell or any one else to prove the origin of a single case of yellow fever occurring on board a ship which has left this port, which was due to the impurity of the river water taken from a point adjacent thereto.   On the contrary, our merchant marine will bear testimony to a preference for the Savannah River water over any that can be supplied to them at other ports generally."

In a paper on "Marine Hygiene, and the Prevention of Epidemic Diseases by Commerce," Dr. A. N. Bell says: " A ship should be, and if well constructed and properly ventilated is, the most healthy of all abodes; yet we are constantly appalled by the rise, spread, and fatality of diseases on board, which even the most ordinary application of intelligence and care would wholly avert.   Persons on board ship are more at the mercy of those who regulate and control it than are the criminals of a penitentiary to a turnkey.

." A vessel may be built and sent to sea with no opening save the hatchway for cargo, so that when it is closed there is no light nor means of renewing the air.   The forecastle, where the sailors sleep, in small vessels especially, is rarely even high enough to admit of an erect posture; and from its position, even in moderate weather, it is frequently necessary to close it, while with the greatest care it is usually subject to being wet.

"Many of our coasting and West Indian traders and steamboats—and some of these latter of palatial-like dimensions—allow of less than fifty cubic feet of air-space to the individual, for the sleeping apartments of their employees. The bunks or shelves, into or upon which the sailors and hands of these classes of vessels are stowed, are rarely or never provided with any opening save the narrow hatchways leading to them. They are the stow-holes for every kind of rubbish, rarely or never cleansed, and probably the most prolific of all the sources of typhus fever in and about New York.

"The writer distinctly calls to mind a visit he once paid to an American whaler brig, in a foreign port, to see the captain, who was ill, where twelve out of a crew of nineteen men had died of ship fever, from these causes. While he was abusing the filthiness of the town and the sickliness of the port, the fore-castle of his vessel would have been disgraceful for a pig-sty."*

Is it necessary to distinguish between Infected and Non-Infected Vessels?

The classification of ships according to sanitary rule demands inspection by a competent officer; and is it not quite as important to see if human beings are to be decimated by foul air and disease, as it is to inspect boilers for protection against explosions?

The Board of Experts stated in their conclusions "that only a small percentage of vessels and boats, coming from infected places, are themselves infected. The discovery of a method which would enable us to discriminate between infected and uninfected vessels, without placing in jeopardy the lives of human beings, would be of inestimable value."

It is not at all likely that any surer test of infection will be found than the history of the ship, if trustworthy, the actual presence of sick on board, and the knowledge of the port whence the vessel has sailed, or of the part of the intra-tropical seas she has crossed. No harm can come, until vigorous inspection and disinfection before and after unloading ships in summer are in practice, from treating all ships as infected, entering the United States from the West Indian Islands, Central America, or Brazil. They should all be disinfected, and the tax for this, trifling as it would be in proportion to the advantages derived from

* "Transactions of the Medical Society of the State of New York for the year 1864," p. 384.

it, might be borne in part by the governments and in part by the ship-owners. We know that steamers are more safe than sailing vessels, and that the soundness and age of a ship and the nature of her cargo enable us to judge of the likelihood of yellow-fever infection.

The loss to the commercial world, the loss to the countries invaded by yellow fever, is so constant and so large, that no outlay demanded for the most absolute prevention of the disease need be a cause of difficulty. The cost of navies for aggressive and defensive purposes is not grudged. Why should a charge, which in comparison to that of navies would be exceedingly small, be refused for the final extirpation of a cruel pestilence? Votes of three millions sterling at a time are agreed to in the House of Commons for an expedition to thrash savages. How much better would such money not be applied to such a purpose as I propose? But sanitary measures are cheaper than armed invasions; and moreover, in the long run, they pay. This may recommend them to merchants, legislators, and statesmen.

Before entering into any details as to the production and operation of cold, or the design of my refrigerating ship, it is well to summarize so far that, whatever measures of disinfection may ultimately be adopted, they should be thorough, conducted under competent and independent officials, not dependent for their berth or salary on ship-owners; and that within certain lines of trade, so long as yellow fever is a disease of the Atlantic Ocean, ships should be disinfected, both before loading to proceed on a voyage, and after unloading, when at their ports of destination. Moreover, if my views are correct, no ship should leave a port, nor be allowed in another, without paying a very heavy penalty, or being liable to prolonged detention, unless some effectual and approved method of constant and positive ventilation has been adopted.

I have lived to see national measures enforced for the prevention of disease which were formerly deemed impossible, and spoken of as unwarrantable interferences with trade interests. I have personally suffered for antagonizing wealthy and influential traders. I now know that, sooner or later, the public insists on life-saving measures, and all it asks is to be led, in its processes and expenditures, by enlightened and upright counselors.

Congress has appropriated money for a first refrigerating ship. If that one is built, I am sure its efficacy will lead to others being placed at foreign as well as home ports, and in some cases machines being erected on land to accomplish a similar purpose.

Cold the Natural Antidote of Yellow Fever.

It has been said that a fundamental difference between the poison of yellow fever and of paludal fever is that the former ceases to have any action ; it is discharged from the system of the person diseased, and dies. Not so with malarial poison, which accumulates, and is thrown off, leaving a latent remnant to renew the periodic attack.

Yellow-fever poison is destroyed by severe frost whenever or wherever it is not protected from it, so that it is exterminated by this means after epidemics, and can only be renewed by reimportation. There are conditions under which it hibernates or lives through a mild winter ; but it is in the folds of retentive clothing, in the interior of frost-proof houses, in dormant centers of decay or putrefaction, beyond the reach of a temperature of 32° F. ; and possibly may it be preserved in a grave.

But, according to my inquiries, the mass of organic molecules floating in the air are precipitated and congealed by cold at the simple temperature of a white frost, and are as effectually disposed of, in all probability, as if a putrid piece of flesh were boiled in a hermetically sealed bottle for six hours. For long, and indeed until quite recently, I hesitated to commit myself unconditionally to this view. I felt that low temperature might scotch and not finally destroy the poison. To-day I agree with Professor Carpenter that "low temperature is the only agency that can be relied on safely to destroy the infection of the disease," and for the following reasons :

1. The abstraction of heat is a simple mechanical process, which can be secured without limit. The lowest temperature may be maintained, if necessary, for a practically indefinite time.

2. Air cooled can be used in enormous volumes to sweep out dwellings and ships, to dry and freeze, to sweep away all products of decay, and exert a lasting antiseptic action.

3. I regard the operation of cold in a yellow-fever ship in

the same light that I would regard the forced blowing of cold air through a close chamber containing putrid meats. Of course it is not proposed to forego the removal of all centers of decay and foulness; but by draughts of intensely cold dry air, all putrefaction can be stopped and bodies mummified as if by heat. It is only a question of time and opportunity.

A quantity of pestilential matter may be wrapped in dry clothing, curtains, and blankets, and preserved, as it is alleged has been done in Memphis; just as sour hams may be housed from one year to another, and when summer returns expedite the decay of sound hams stored with them. For this reason scrubbing, washing, and either burning or disinfecting by chemicals, must come to our aid, whether we use cold or not. If a man is told to wash himself, he is also told to put on clean clothes. When we attempt to purify a ship or a house, we must purify thoroughly; and, with it all, we may occasionally be defeated in removing every source of infection.

How to Produce and Apply the Cold.

It would be foreign to my purpose, in this monograph, to describe freezing machines or the details of the refrigerating ship. In a communication addressed by me to Congress on the 3d of February, 1879, I wrote as follows:

To the Committees of the Senate and House of Representatives of the Congress of the United States, upon the subject of Epidemic Diseases.

Experience proves that the poison of yellow fever can with difficulty be reached by chemical antiseptics and disinfectants, more especially in the lower portions of laden ships, where it most frequently lurks. It may be scalded out by intensely hot steam, but this is destructive to the cargo and to the ship's fittings; more or less it is harmful even to the hull, since a high temperature distributed unequally causes irregular expansion, buckling of plates, and starting of rivets.

Considerable experience in sanitary work led me to suggest to Surgeon-General Woodworth that a handy steamer, of adequate strength and form, might be built to carry a powerful refrigerating machine and a standing store of cooled uncongealable liquids, whereby, at short notice, any ship might be effectually subjected in every part to any desired temperature, above or below the freezing-point of water.

Dr. Woodworth most readily seconded my suggestion, by causing an experienced draughtsman at the Navy Yard to be detailed and coöperate with my own engineer, Mr. W. E. Sudlow, to make the necessary designs. It was first my duty to establish the basis of this work, and I determined that, in dealing with the largest ships usually entering such a port as New Orleans, it would be necessary to provide sufficient power to produce an effect equal to the melting of from fifty to one hundred tons of ice, as fast as refrigerated air or liquids could be distributed in the infected vessel, viz., within an hour or two. Whereas the average size of merchant vessels entering the port of New Orleans is under 800 tons, I have deemed it wise to calculate the machinery for the disinfection of the largest vessels usually entering that port, viz., of 1,500 tons. Even these can be so rapidly refrigerated, by the contrivances designed, that it would only require a few more hours' work to deal with much larger steamers.

Disinfection of Air.

The Board of Experts, authorized by Congress, to investigate the yellow-fever epidemic, have declared that "in the dissemination of yellow fever atmospheric air is the usual medium through which the infection is received in the human system." It is therefore of paramount importance to deal effectually with the air confined in the ship. We have provided for the complete circulation, washing, and cooling, *to any desired temperature*, of the entire volume of air in a 1,500-ton ship, in five minutes. The blower draws and delivers 26,400 cubic feet of air per minute, whereas the ship holds, empty, 150,000 cubic feet of air.

No part of the air can remain stagnant or quiescent, for no part of a ship is made so air-tight that provision can not be made to exhaust effectually by a suitable apparatus.

Disinfection of the Bilge.

Pumps of adequate size will clear out the bilge as much as possible. But the difficulty, especially in wooden ships, is the adherent organic matter, which penetrates the wood and remains there a constant source of danger, discharging into every fresh influx of air, or new cargo, more and more of the infecting material.

To meet this, a strong solution of chloride of magnesium will be pumped into the bilge; and wherever necessary, under a high pressure, by means of a hose, a current of this artificial "pickle" will be driven into the crevices and pores of the wood. Chlorides used in this way have a preservative influence on the wood, and effectually control the putrefactive or fermentative changes which the fetor indicates as concomitants of the contagious principle. The solution thus forced may be at 20° or 30° below zero of Fahrenheit, if necessary; and on the assumption of the Board of Experts, that "the yellow-fever poison is not able to withstand the influence of frost," it is not difficult to maintain the frozen liquids in circulation, in contact with the contaminated parts of a ship, so long as to render innocuous and even destroy such poison.

So far as the purification of the bilge is concerned, the action of deliquescent chlorides and glycerine, such as I employ in my refrigerating machinery, has an effect independent of temperature. My own experiments with inoculable virus prove that these agents destroy the reproductive power of animal or certain vegetable organisms and poisons by their affinity for water. They shrivel and modify the chemical constitution of any virus, and thus affect its vitality.

We therefore have two weapons in our hands, and I can with great confidence declare that the most difficult branch of a ship's disinfection, viz., purifying the bilge, is the easiest and most certain in the way proposed.

Disinfection of Cargo.

Again, referring to the conclusions of the Board of Experts, we find at page 27 that artificial cold may "prove more efficient in destroying the infection of clothing, baggage, and many kinds of goods, than any other artificial means which can be employed."

Again, they say, "Atmospheric air is an efficient, as well as a generally applicable, disinfecting agent, both as respects yellow fever and cholera."

It therefore follows that, according to the kind of cargo imported by any ship, a more or less reduced temperature of air-currents driven through it may afford protection ; but the system of "cold storage," widely adopted even in Northern cities, is available at quarantine grounds, and a ship, relieved of its burden, may proceed again on its voyage, practically without detention.

The cargoes of fruit, which so commonly carry infection, may be benefited by a reduction to 32° short of freezing the fruit. Every Southern city should have its cold stores, and in this way the produce meets with a steadier market, and the sacrifice due to "gluts" may be permanently averted.

Refrigerating Machinery.

The steamer designed is to be built of steel, and stanch enough to bear the incessant strain and work incidental to an active service such as I have contemplated.

Apart from the compound engine and propeller, to drive the boat, it will hold a thermo-glacial engine of my latest design, which is practically independent of the temperature of condensing water. The abstraction of heat is equivalent to one ton of ice for every 425 pounds of pure ammonia gas, alternately liquefied and evaporated.

Unlike the freezing machinery hitherto used, especially for the supply of ice in New Orleans, whatever the heat of the weather or of the waters of the Gulf of Mexico, my thermo-glacial engine will produce the same amount of refrigerating effect, and will abstract 30° of heat every time the 26,400 cubic feet of air above mentioned flows through our cooler.

Store of Cold.

In order to concentrate the power of the machine, it will always have from forty to fifty tons of chloride of magnesium solution in the ballast

tanks of the steamer, cooled down below zero of Fahrenheit. This reserve will enable us at short notice to overwhelm by cold every part of
the foulest ship, without danger of interference by surrounding temperatures.

It may be necessary to freeze more than one moderate-sized ship per
day, and by this process the freezing machinery can be kept at work night
as well as day, and be constantly in proper condition to meet the requirements of trade.

Cost of Refrigeration.

Making allowance for the most skillful sanitary and engineering super-
intendence, this steamer will cost about $2,000 per month, to be kept in
operation. Calculating that thirty ships only are purified within the
month, the cost of disinfection will be under $70 per vessel. What is this
compared to the cost of protracted quarantine?

Writing on the 22d of April to the National Board of
Health, I added : " That a ship should be used as a floating
sanitary laboratory or workshop in the way suggested is entirely
novel, and the points which led me to its adoption were as follows :

" 1. An infected ship is often in distress, owing to the sickness of its crew or other accidents, so liable to occur when officers and men are mentally and physically disabled. Whether
in distress or not, a pilot has to board her, and thus a constant
link is established between floating centers of infection and
land. In addition to a pilot, a tug is often needed, and the tug
with its crew constitutes a second source of danger. The disinfecting or refrigerating ship can therefore serve with manifest
advantage as a pilot-boat, a tug, and a purifying vessel, carrying all the appliances to minister to the comfort of the sick and
the protection of human lives.

" 2. A ship is available with its disinfecting or refrigerating
machinery over a wide area near a quarantine station, and, as in
the case of New Orleans, may steam down to the bay and prevent the towing of infected vessels past Port Eads and between
fairly populous lands, before it can possibly be taken charge of
by sanitary officers in quarantine.

" 3. The disinfecting ship may have to deal with the cargo
of an infected vessel first, disinfect this and land it, and then
proceed to disinfect the ship by refrigeration or otherwise.

" 4. The refrigerating machinery on board the disinfecting

vessel can be utilized at quarantine while lying alongside; and, if desired by the National Board of Health, I will supply designs for the most efficient form of cold disinfecting chamber, to be erected on land, and maintained at a low temperature by the refrigerating ship."

" In addition to refrigeration, I propose to provide the following means:

" a. Deliquescent chlorides, and especially the chloride of magnesium, to remove from the air all molecular matter and moisture. The air will be thus rendered *absolutely* and *optically* *pure*. The bilge, after complete pumping and washing, will likewise be purified by chloride of magnesium solution at low temperature. Muddy ballast, etc., can be rendered innocuous by such treatment.

" b. A close chamber, to be raised to any desired temperature for disinfecting purposes, or in which fumigation may be thoroughly carried out, to purify loose and movable objects, clothing, etc.

" c. An entirely new system of temporarily substituting for the air of a ship antiseptic gases, measured, weighed, and introduced as liquids in the most inaccessible parts, and which, passing spontaneously from the state of liquid to that of gas, will permeate everywhere, and, if necessary, be made to penetrate under pressure. Then fumigation will not be simple chance work.

" d. Electrical apparatus for lighting and safely inspecting every part of an infected vessel, or for conducting work during the night.

" Electrical indicators for automatically registering the temperatures pervading every part of a ship during the process of purification. These self-registering indicators will serve to prove when and how a ship has been purified at a home or foreign port.

" e. Chemical, spectroscopic, and microscopic apparatus of entirely novel character, to determine at any moment the condition of the atmosphere; means of collecting and examining air and all atmospheric impurities.

" f. Medicated or disinfected baths, or chamber in which convalescent people, or persons exposed to contagion, may be purified before landing.

"*g.* Apparatus for superheating, and with an entirely novel method of so regulating superheated steam, if ever used in disinfecting a ship which has been stripped for the purpose, as to maintain safe temperatures and prevent firing or other inconvenience."

I have planned a variety of methods, which I know from experience to be satisfactory, whereby inclosed spaces can be purified; and it must always be a matter of importance to discover the most readily applied and most economical that are compatible with a sure result.

The National Board of Health, and under it a Board of Naval Engineers, have approved of my plans and suggestions, and recommended them for practical execution. It is best, therefore, to enforce the lessons to which I have devoted these pages by direct experiment; and extracts from the report of the Board of Naval Engineers may prove instructive and encouraging to those who are disposed to follow my advice.

Report of the Naval Engineers on the Refrigerating Ship.

That Board published their report on the 12th of June, after six weeks' close and searching investigation. They say:

The problem of refrigerating a vessel carrying a miscellaneous cargo differs materially from anything heretofore attempted, and the data furnished by machines employed in ordinary refrigerating processes can not be applied to the solution of the present problem without material modifications.

The conditions under which the work has to be done will vary greatly with the size of the ship to be refrigerated, the material of the hull, the character of the cargo with which she is laden, the temperatures of the surrounding water and air, and the temperature of refrigeration to be maintained in the vessel.

By reference to the steam-logs of several United States naval vessels, it was found that the mean temperature of the water is about 84° F. Under the above conditions, the temperature of the vessel must be reduced to at least the freezing-point of water in eight hours, and the machinery must be capable of maintaining a temperature of 0° F. for an indefinite period of time, while the vessel is being unloaded, cleaned, and disinfected.

A vessel having a register tonnage of 1,500 tons has a capacity of 150,000 cubic feet. Taking the mean of the dimensions of several vessels of this capacity, but of different models, we find that the immersed surface of such a vessel when laden is 10,800 square feet; the area of the sides when laden is 5,200 square feet; the immersed surface when empty is 8,500 square feet; the area of the sides when empty is 7,500 square feet; the area of the deck is 6,600 square feet.

The total weight of the hull proper of an iron vessel of the given capacity is 730 tons. Deducting 50 tons from this weight for the rail and other upper works, the stem, stem-post, etc., we get for the weight of the hull actually inclosing the space which is to be refrigerated 680 tons.

The work to be done during the first eight hours has been calculated in British units of heat, being the sum of the following quantities:

1. The units of heat absorbed in reducing the temperature of the air in the vessel from 84° to 32° F.

It has been assumed that the cargo actually occupies 70 per cent. of the cubical capacity of the vessel, leaving 45,000 cubic feet of air. Since the volume of air is reduced 10 per cent. in lowering its temperature from 84° to 32° F., 4,500 additional cubic feet will enter the vessel during refrigeration, making the total quantity 49,500 cubic feet.

2. The units of heat absorbed in condensing the watery vapor contained in 49,500 cubic feet of air at 84° F. when in a state of saturation, and in lowering the temperature to 32° F.

No allowance has been made for the heat set free by the freezing of the water which remains in the vessel after it has been pumped as dry as possible.

3. The units of heat absorbed in reducing the iron hull, weighing 680 tons, from a temperature of 84° to 32° F.

A wooden hull is about 10 per cent. heavier than an iron hull, and the specific heat of wood is far greater than that of iron; but, on account of the relatively low thermal conductivity of wood, the final temperature of refrigeration would be reached, and could be maintained a long time within the vessel, before the heat would be abstracted from the whole mass of wood.

4. The quantity of heat absorbed from the cargo is assumed to be equal to the quantity of heat absorbed from the hull.

This assumption appears to furnish a safe average. In case the cargo consisted of metals, the whole mass would give up its heat; while in the case of a closely packed material having a higher specific heat, but lower thermal conductivity, relatively little heat would be absorbed.

5. The quantity of heat transmitted through the immersed hull from the water, being calculated by the following formula of Péclet, adapted by Rankine to British units of measure:

$$q = \frac{t - t_1}{\dfrac{1}{a\,(1 + B\,(t - t_1))} + \rho\kappa} \qquad (1),$$

where q is equal to the number of units of heat transmitted per hour through one square foot of surface of a plate x inches in thickness. The aver-

12

age thickness of iron hull, considered as a uniform solid wall, separating the cold air inside the vessel from the warm water surrounding it, has been taken as being equal to three fourths of an inch. The value of t, representing the temperature of the water in motion, is constant, viz., 84° F. The value of t_1, representing the temperature of the refrigerated air, varies between 84° F. at the commencement of refrigeration and 32° F. at the expiration of eight hours. The transmission of heat has been calculated for a mean difference of temperature of 26° F. for eight hours.

6. The quantity of heat transmitted by contact of the warm outside air with the sides and the deck of the vessel. This quantity is calculated by the following formula of Dulong and Petit, as modified by Péclet, viz.: $C = 1136 \vartheta^{1233}$ (2), in which C is the number of *calories* (or French units of heat) transmitted per square *mètre* per hour, when ϑ represents the difference of temperature in degrees centigrade. (A *calorie* is transformed into British units of heat by multiplying it by 3·96832.)

Since the transfer of heat through the deck and the sides of the vessel may be prevented to a great extent by covering them with blankets, mattresses, sails, etc., only one tenth of the quantity found by equation (2) has been taken into account.

As only a rough approximation of the quantity of heat thus transmitted was attempted, the difference of temperature has been taken as being one half of the final difference during the whole eight hours; that is to say, $\vartheta = 14·44°$ C.

The absorption of heat from and the transmission of heat through the hull form so large a portion of the total heat to be absorbed in the case of an iron vessel, while these quantities would be relatively small with a wooden hull, that a machine capable of refrigerating an iron vessel will be amply sufficient for a wooden one.

The number of units of heat to be abstracted per hour by the machine in order to maintain the temperature of the vessel at 0° F. has been calculated from formulas (1) and (2), assuming that the temperature of the vessel is 0° F., and that the areas of the immersed hull and of the sides are those obtaining when the vessel is empty.

Units of heat to be abstracted in the first eight hours.

Units of heat abstracted in reducing temperature of 49,500 cubic feet of dry air from 84° to 32° F.	44,645
Units of heat abstracted in reducing the vapor contained in the air, when saturated, to water of 32° F.	98,722
Units of heat abstracted in reducing iron hull from 84° to 32° F.	8,696,863
Units of heat abstracted from cargo (estimated)	8,696,863
Units of heat transmitted from water through iron hull	3,868,300
Units of heat transmitted from air through protected iron hull	106,423
Total units of heat to be abstracted during first eight hours	21,511,816
Average work to be done per hour	2,688,977

Units of heat to be abstracted per hour in order to maintain the temperature of the vessel at 0° F.

Units of heat transmitted per hour from water at 84° F. through iron
hull having an interior temperature of 0° F.............................. 1,469,000

Units of heat transmitted per hour from air at 84° F. through protected
iron hull having an interior temperature of 0° F.................. 69,000
                                                                   ─────────
     Total... 1,538,000

Although the work required to be done per hour in reducing the temperature from 84° to 32° F., during the first eight hours, is much greater than that required to maintain the temperature of the vessel at 0° F., a machine capable of doing the latter amount of work is sufficient if a store of some cold medium is provided to be used in aiding the machine, during the first eight hours, in reducing the temperature to 32° F.

The transmission of heat through the hull increases so rapidly with a reduction of the interior temperature, that a machine capable of abstracting the heat transmitted, when the interior temperature is 0° F., will be sufficient to abstract any additional heat introduced during the hoisting out of cargo, etc., and gradually to reduce the interior temperature from 32° F. till the limit of its capacity at 0° F. is reached.

Since the work of refrigeration is to be done by introducing cold air into the vessel under treatment, it is convenient to express this work in terms of cubic feet of air. The work of abstracting 1,538,000 units of heat is equivalent to lowering 75,000,000 cubic feet of air at zero of Fahrenheit 1°, or 1,000,000 cubic feet of air 75°.

In order to maintain the temperature of the vessel at zero of Fahrenheit, the refrigerating air entering the vessel must have a lower temperature, so as to absorb the heat transmitted through the hull; the volume of the entering air has to be increased in the direct ratio in which the difference of temperature of the entering air and the hull is decreased. It appears impracticable, for several reasons, to let the air enter the vessel at a lower temperature than about −25° F., and 3,000,000 cubic feet of air of this temperature would be required per hour to maintain the temperature of the vessel at zero of Fahrenheit; and to maintain this temperature uniformly at every point, the above quantity would have to be evenly distributed over and brought into direct contact with the surface of the hull. This, however, is manifestly impracticable, and great differences of temperature must consequently always exist within the vessel, and a large quantity of the refrigerating air will pass through the vessel without producing any useful effect.

We must, however, assume that, in order to maintain the required temperature, the machine must be capable of reducing 3,000,000 cubic feet of air from zero to −25° F. per hour.

In the plans submitted, the refrigeration is to be effected by the compression and expansion of air, or by the evaporation of some volatile liquid.

The cooling of air by its own compression and expansion recommends itself at first sight, on account of the apparent simplicity and effectiveness of the process.

The machinery required for this purpose is simple in its construction, consisting of an ordinary steam-engine and boiler, cylinders for compressing and expanding the air, and an intermediate reservoir, in which the compressed air is cooled to, approximately, its initial temperature, with pumps for circulating the cooling water. The ordinary running expenses of this machinery are represented by the consumption of fuel in the boiler, and the expenditures for attendance, lubricants, and the renewal of valves and packings. With coal in the bunkers, the machinery may at all times be ready for operation. No difficulty exists in securing at any time, for its manipulation, persons thoroughly familiar with all its parts, experienced in working compressed-air engines, and competent to attend to any repairs and adjustments which may be required.

The temperature of the air may be reduced by this process probably to a lower point than by any other applicable on a large scale.

The principal advantage connected with the production of a very low temperature in the air-machine, by using a high degree of compression and expansion, consists in the fact that a smaller volume of air is required for the absorption of a given quantity of heat, and consequently smaller cylinders may be employed; on the other hand, the excess of power expended in compression, over that recovered by the expansion of the cooled air, increases with the degree of compression, while the loss of power, due to friction and leaks, and the irregularity of the resistance, increases likewise with the higher final pressure.

In order to produce the greatest effect with refrigerating machines working with compressed air, it is necessary, first, to compress the air at a constant temperature; second, to avoid exceeding the lower temperature strictly necessary for producing the desired frigorific effect.

The abstraction of the heat generated during compression is effected in the simplest and most efficient manner by injecting water into the compressing cylinder; surface cooling alone has been abandoned in air-machines. The several devices presented, for increasing the cooling surface in the compressor, will be found inefficient or impracticable.

The final temperature of the compressed air must depend on that of the cooling water. It is, however, possible to utilize to some extent the unavoidable waste of refrigerated material in further reducing this temperature.

It is important, for a twofold purpose, that the air should enter the expanding cylinder as dry as possible: First, the heat set free by the condensation and congelation of the watery vapor would raise the temperature of the expanded air; secondly, the formation of ice in the expanding cylinder would tend to choke up the ports and passages, and increase greatly the friction of the moving parts.

In order to reduce the temperature of saturated air from $84°$ to $-25°$ F.

(making allowance for the rise of temperature due to the congelation of the vapor) by expanding it to a pressure of 15·5 pounds per square inch, its initial pressure must be 48·73 pounds per square inch. This pressure can be attained, without difficulty, by compression in a single cylinder.

In some of the submitted plans it is proposed to compress the air in stages up to from 80 to 120 pounds per square inch. This would entail a useless waste of power and unnecessary multiplication of machinery.

Most of the proposed air-machines are so ill-proportioned, and the efficiency claimed for them is so greatly overrated, that it has been considered instructive to calculate the principal proportions and the power required for air-machines, capable of doing the work of refrigeration required under favorable conditions, in order to present a standard for comparing the efficiency of the various plans and methods of refrigeration.

The maximum speed of large air-machines should not exceed, according to good authorities, 360 feet per minute, and at this speed the mass of air discharged would be equal to about 80 per cent. of the space swept through by the piston. To discharge 3,000,000 cubic feet of air per hour would require, at this rate, eight compressing cylinders, each 63 inches in diameter, and having 6 feet stroke of piston, the engines making 30 double strokes per minute. Taking the atmospheric pressure at 14·7 pounds, the initial pressure in the compressing cylinder would be about 13·5 pounds.

	Pounds.
In compressing this to 50 pounds per square inch at a constant temperature, the mean unbalanced pressure per square inch of piston is	17,658
And the mean unbalanced pressure on each piston	55,044
The friction of the leather packing-rings of each piston would be about	2,517
Total resistance of each piston	57,561

To discharge the same weight of air at a temperature of −25° F. and a pressure of 15·5 pounds per square inch, there should be eight expanding cylinders of 47 inches × 72 inches. Assuming that the air enters in a state of saturation, at a pressure of 49 pounds and at a temperature of 84° F., and that it expands sufficiently to lower its temperature to −25° F. (making allowance for the rise of temperature due to the heat set free from the congealed vapor), the unbalanced mean pressure per square inch of piston is 21·26 pounds; the unbalanced mean pressure on each piston, 36,886 pounds; the friction of the leather packing-rings of each piston would be about 2,258 pounds; effective pressure on piston, 34,628 pounds; difference between resistance in each compressing cylinder and effective pressure in each expanding cylinder, 22,933 pounds; indicated horse-power of steam-engine required to discharge 3,000,000 cubic feet of air under the above conditions, assuming the efficiency of the engine to be 80 per cent. of the total power, 2,500.

This enormous expenditure of power is due to the fact that the work to be done by an air-machine depends on the volume of air discharged and on the respective temperatures during compression and expansion, without reference to the quantity of heat abstracted during refrigeration. The temperatures range in the present case between 84° F., the temperature of the injection water, and —25° F., the temperature of the expanded air. By utilizing the cold air leaving the vessel at 0° F. in reducing the temperature of the injection water, the power required for refrigeration may be reduced to some extent.

The disadvantages under which all refrigerating air-machines labor are concisely stated by M. A. Terquem in a paper read before the French Academy of Sciences in 1877. He says :

"But even under the best conditions it does not seem possible for the cold-air machines to compete with machines worked with volatile liquids. They present, in fact, the same inconveniences as the hot-air motors. These inconveniences are as follows:

"1. The necessity of using large dimensions in order to obtain small effects, on account of the low density and the low specific heat of air.

"2. The passive resistances due to these large dimensions and to the use of two cylinders; these resistances being proportional to the sum of their energies, the work to be expended being proportional to the difference of the same.

"3. The inability of these machines to accommodate themselves to different degrees of refrigeration, the dimensions of the cylinders having to be proportioned for the temperatures T and T₁, between which the machine has to function."

Among the volatile liquids, or liquefiable gases, used in refrigerating machines, there are several, like ether, chymogene, and sulphurous acid, which should not be used on board of a steamer on account of their explosive or poisonous character.

These objections do not exist in the case of ammonia, while its efficiency as a cooling agent is eminently high on account of the low temperature at which it evaporates under ordinary pressures, and the large quantity of heat which becomes latent during its evaporation. Difficulties experienced at first in using such agencies have been overcome by making the compressing pumps single-acting, thereby preventing leakage through stuffing-boxes, by introducing various devices equally effective for avoiding all loss from clearance at end of stroke, and by adopting such forms of condensers and refrigerators as to insure a rapid transmission and absorption of heat.

Confining ourselves to the consideration of machines working with anhydrous ammonia, it becomes apparent at once that there exist still many grave difficulties connected with the use of this gas. Its great affinity for many substances, in common use in other motive machinery, makes necessary a careful selection of the material with which it may come in contact.

At the mean temperature of our Southern ports, this gas has to be worked on very high pressures. The great differences of pressure existing in the machine require an unusual degree of exactness in the fitting of the working parts. The necessity of making many of the joints rigid by soldering increases the liability to leakage, in consequence of unequal expansion from differences of temperature. The introduction of this agent is of relatively recent date, and it may be difficult to procure at all times the services of persons thoroughly familiar with its character, and competent to manipulate large and complex machinery in which it is used. The escape of this gas in small masses causes annoyance; in large masses it might be dangerous, by causing suffocation.

Hence ammonia-machines must possess very decided advantages, if they are to be selected instead of the simple but bulky and expensive air-machines for purposes of refrigeration.

The Board has experienced much difficulty in securing reliable data regarding certain physical properties of ammonia necessary for calculations of the power developed in its compression and expansion ; and the following calculations are to be accepted only in so far as the laws governing the compression of permanent gases are applicable to ammoniacal gas.

In order to reduce the temperature of 3,000,000 cubic feet of air from 0° to —25° F., the machine has to work under the following conditions. Pictet states that in his ice-machines, working with sulphurous acid, there must be about 20° F. difference of temperature at both extremes of the process, between the gas and the water used for congelation on the one hand, or for removal of excess of heat on the other. The same differences have to exist in the ammonia-machine, except that, for refrigerating the air, this difference has to be somewhat greater. Hence we may assume that, since the temperature of the cooling water is 84° F., the temperature of the gas in the condenser will be 104° F., and that the temperature of the gas in the refrigerator has to be about —50° F. in order to cool the air to —25° F.

The pressures of saturated vapor of ammonia, corresponding to 104° F. and 25° F., are 227 pounds and 8·2 pounds per square inch, respectively. Allowing for resistance in passages, and the excess of pressure required for opening the induction-valve, the gas will enter the pumps at a pressure of about 6 pounds per square inch.

Pounds of liquid ammonia, of 104° F., which has to be evaporated under the pressure of 8·2 pounds in order to absorb 1,538,000 units of heat per hour.. 1,880

Pounds of ammonia which have to be evaporated per minute..... ...... 31·33

Capacity of pumps required for exhausting this weight of gas at a pressure of 6 pounds per square inch, expressed in cubic feet, displaced by piston per minute... ... 1,345·8

Work to be done in compressing 31·33 pounds of gas, having an initial pressure of 6 pounds and discharging it at a final pressure of 227 pounds per square inch, expressed in foot-pounds, per minute........ 6,614,230

Indicated horse-power of steam-engine to do this work, assuming that the work done in overcoming friction of load and resistance of engines proper is equal to 40 per cent. of the useful work...................................... 280

Indicated horse-power required to run pumps for circulating water and refrigerating medium, and fan-blowers for furnishing 3,000,000 cubic feet of air per hour.. 100

Total indicated horse-power of engines....................... 380

It will be seen that, to produce the same frigorific effect under the given conditions, the ammonia-machine requires less than one sixth of the power required for working an air-machine. This increase of power means not only increase of running expenses, but increased bulk, weight, and cost of machinery, and consequently, in the present case, increased size and cost of vessel which is to carry the refrigerating apparatus. Two ammonia-pumps of moderate dimensions, worked by a pair of ordinary steam-engines, and connected with tubular condensers and refrigerators, two small pumps for circulating the condensing water and refrigerating medium, and two or three large fan-blowers, could do the work under consideration, for which, on the other hand, eight ponderous air-machines, each consisting of two air-cylinders, besides the necessary steam-cylinders, would be required.

In all machines of this kind, when used on board of a vessel, the unavoidable fly-wheels form a very objectionable feature ; although the variations in resistance are far greater in the ammonia- than in the air-machines, the smaller size of the former reduces greatly the inconveniences resulting therefrom.

It is a decided advantage that with the ammonia-machine the quantity of heat abstracted and the volume of air furnished in a unit of time can be varied at will independent of each other.

The ammonia-machine is likewise free from the great waste of power which is unavoidable in air-machines, when working with air having an initial temperature much lower than that of the cooling water.

Much difficulty has been experienced in cooling air by direct contact with the intensely cold tubes of the refrigerator, because the vapor present in the air would congeal instantly, and, coating the tubes, would diminish greatly their heat-conducting power. This difficulty is obviated by using some liquid as a refrigerating medium which does not congeal at the temperatures obtaining in the refrigerator. This liquid falls in a shower into a chamber, through which the air which is to be cooled is forced, and, after abstracting its heat, is returned to the refrigerator.

After disposing of the various plans which failed to meet the requirements of the Board, the following observations are made concerning my plans and appliances :

The characteristic features of this process of refrigeration are the expansion of the compressed gas ; the use of rotary pumps for the compression and expansion of the gas; the cooling of the refrigerating air by direct contact with a shower of a cold uncongealable liquid; and the storing of a supply of this cold liquid in insulated tanks.

The expansion of the compressed gas in a working cylinder is a feature not presented in the other plans submitted using ammoniacal gas, and reduces the expenditure of power very materially.

In the first place, the pressure in the condenser may be reduced approximately to that of saturated vapor at the temperature of the cooling water —that is to say, in the present case, from 227 pounds to nearly 142 pounds; and the work of compression will be proportionately lessened.

Secondly, part of the work of compression is utilized by the expansion of the gas. Consequently a material reduction would have to be made from the result of the calculation heretofore presented, in the estimate of the power required to work this machinery. In the plan proposed, provision is made for pumping the liquid ammonia from the condenser to a boiler, where it is to be evaporated under a pressure of about 300 pounds per square inch, to be then used in the expanding cylinders.

Although the use of ammonia as a motive power may be more economical than that of steam, the Board are of the opinion that the difficulties to be apprehended from the increased complication of the operation, and from the additional valves, joints, and pumps, outweigh the possible economical advantages, and that the expansion-cylinders should draw their supply of gas directly from the condenser.

The use of rotary pumps for compression of gases to a high tension has a decided advantage over that of reciprocating pumps, in avoiding all shocks due to sudden changes of motion under great variation of pressure. Difficulties that have been encountered with pumps of this type, especially on account of leakage, have been successfully overcome in the construction of the pump shown in the plan. It is likewise proposed to use rotary engines of identical construction for the steam-motor, to be used both for propelling the vessel and for driving the refrigerating machinery.

The use of the same engines for the double purpose of propelling the vessel and driving the refrigerating machinery possesses the great advantage of simplicity and economy.

The so-called uncongealable liquid proposed is a solution of chloride of magnesium in water, with a small quantity of glycerine, which solidifies at about the freezing-point of mercury. This solution is forced by an independent reciprocating pump through the refrigerator, which consists of a series of tubes of unequal diameters.

The larger tubes are rigidly secured at both ends in fixed tube-heads; the smaller, which pass through the larger ones, are fitted in removable heads and packed at both ends. The ammonia circulates through the annular spaces thus formed, and the liquid to be cooled through the smaller and around the larger tubes. This arrangement presents a very large and effective cooling surface in a small space.

A rotary blower is coupled directly to the engine-shaft.

The capacity of the pumps and the engine and boiler power are sufficient for the required work of refrigeration.

The proposed appliances for distilling the anhydrous gas from the ammoniacal liquor, the provisions for storing it, and the various devices for obviating the difficulties arising from the high pressure and the subtle character of the gas, indicate a careful study of the whole subject.

The arrangement of the deck-houses of the vessel proposed to carry the machine shows the same careful consideration of the subject of quarters, bath-rooms, sick-chambers, instrument-room, etc., which may be required in connection with the proposed object of disinfection.

It appears doubtful whether the volume of extremely cold air required for maintaining the temperature of a 1,500-ton iron vessel at zero of Fahrenheit can be so distributed as to be effective in completely absorbing the heat transmitted through the hull; and it has to be assumed that, even after the most careful investigation of such a novel subject, certain unforeseen circumstances may exist which would increase the difficulties of accomplishing the desired object. To provide against such uncertain elements, and at the same time for losses by conduction and decrease of efficiency from wear and tear, it would not be prudent to allow less than 25 per cent. in excess of the calculated capacity; or, the power of the machine being proportioned to the calculated work required in reducing the temperature of a 1,500-ton iron vessel to zero of Fahrenheit, a correspondingly lower actual performance has to be expected.

After a close study of the problem to be solved and a careful examination of all the plans presented, the Board find that the plans submitted by John Gamgee show the most thorough and intelligent consideration of the subject, and promise to do the required work most efficiently and economically.*

Cooling and Ventilating Ships.

"The state of the air in ships," says Dr. D. Boswell Reid, " is much influenced by their construction, the materials of which they are formed, the purposes to which they are applied, the presence or absence of the steam-engine, the climates which they visit, the cargoes which they carry, the number of persons that may be crowded within a given space, the diet that may be provided, and the discipline and usages which may be enforced."

Without attempting to forestall observations which will

* For the full report the reader is referred to Senate Document No. 30 of the first session of the XLVIth Congress, of which the foregoing is an abridgment. The Board was composed of Mr. David Smith, Chief Engineer, U. S. N.; Mr. C. R. Roelker and Mr. W. A. H. Allen, Passed Assistant Engineers, U. S. N.; and Mr. W. L. Montanye, Naval Constructor, U. S. N.

shortly be published in a special work on ventilation, I may state that the permanent prevention of yellow fever imperatively demands the supervision of ship construction for the preservation of the seaman's health. I have already explained that the natural conditions on land, which protect human lives in badly built dwellings, are absent in a hide-bound floating structure. Wind-sails to force air into a ship, and cowls or wind-sails to exhaust, are very useful appliances on shipboard; but for all immediate practical purposes a small blower on Mr. John A. Svedberg's patent, such as I have introduced into my refrigerating ship, should be connected with a pipe or pipes at the lowest possible part of a ship, and made to extract foul air constantly.

The primary object, in preventing yellow fever, is to relieve the ship of all stagnant air or putrescible liquids. The bilge-pumps must be used constantly and frequently. Sailing vessels of adequate size would be benefited by carrying a donkey engine and boiler for this pumping and positive ventilation of ships. The engine could also be used to help to load and unload the cargo. Human hands may be made to pump and turn a blower; but the work is very heavy in tropical climates, and the moment sickness begins to disable a crew, such work has to be neglected, to the aggravation of all unhealthy conditions.

High Temperature as a Yellow-Fever Antidote.

I have noticed a tendency among some engineers and medical men to believe more in the efficiency of steam and high temperatures than cold; but a careful consideration of the subject indicates that steam is usually impossible of application, can not fail to injure the ship's fittings, and must leave a vessel, especially if of wood, in a much worse condition than if treated by a dry process.

Steam has proved sufficient to disinfect a man-of-war; but then, simple aëration and thorough cleansing have accomplished the same purpose in favorable cases. If the putrescent matter in a ship, adhering to the sides, penetrating crevices and decaying wood, is of the nature of ordinary putrefactive matter, we have experiments to prove that the direct subjection to a temperature of 212°, for five or six hours, is essential in order to prevent the renewal of life in the putrefying center.

We have, no doubt, to deal with a special product which frost can destroy, but there are indirect advantages in the process which I have from the first projected. It is not at all improbable that when a city is subjected to a frost the gravitation of atmospheric impurities to the soil *en masse* exerts a most powerful influence; for once it is deposited, solid or liquid, and the air is left free, it would require a renewal of high temperatures, such as does not occur after a frost, to set the active elements into motion again, and enable them to infect the air by indefinite reproduction.

When a blower is started with a shower of chloride of magnesium cooling the air, all moisture and organic matter is seized and fixed by the strongly absorbing solution. If we blow fast and long enough, the air is freed of everything that does not rise at low temperatures from the objects included in the vessel; and, having emptied these out, we can treat the residual surface and substances *secundum artem.*

Cold, therefore, enables us to imprison as well as destroy offending matter, and it is not in the least harmful to ship and cargo.

Above all, my plan dries everything completely; and from first to last I must insist on that most dangerous of all atmospheric agents to the crew, as Lévy puts it, viz., humidity. As Dr. Turner wisely says: "The actual humidity of the air on decks at sea, or anywhere else, should never be supplemented by artificial means to render it saturated." Above all, let us bear in mind Captain Cook's experience.

With the ready methods compatible with prompt disinfection of ships in port, it would be a most serious matter to steam and then attempt a drying process. Such measures may be resorted to with naval vessels put out of commission at will, but are certainly not practicable for the mercantile marine.

Fire as a Disinfectant.

The drying of ships by currents of hot air has been adopted, perhaps insufficiently but frequently, since the days of Captain Cook.

A simple fire—not a pleasant companion in the tropics—may be made to draw or drive the air from the bilge and other parts of the vessel; but the Sanitary Council who may have charge of international regulations for the extirpation of yellow fever must remember that any power used to modify the nature of air in ships must be supplemental, or practically used by the employment and construction of general and suitable channels of air-conduction, which will lead to a complete renewal of air in a ship.

Steam-coils, a tubular heater, or any ready method of heating air in a circuit, may, with a capacious blower, be of service, but more to desiccate than to attempt the destruction of the poison by high temperature.

M. L'Apparent, Director of Naval Construction in the French Navy, is mentioned by M. Mélier as having proposed the *gas-flame apparatus*, which painters use to burn paint off wood, whereby to singe the whole interior of a vessel. This *fire douche*, as he calls it, may scorch out the disease under careful supervision; but the process is tedious and needlessly destructive of the ship's internal surfaces.

On this question of ventilating ships by heat, Dr. Reid (p. 361) says:

" *The fire of the galley* has occasionally been converted into a ventilating power; but, so far as I am aware, no general and systematic distribution of communicating channels has accompanied the application of this power. Nothing, however, more certainly tends to improve the atmosphere of a ship than the influence of the galley-fire, as it may be made to serve the same purpose as a shaft; but the objections to sustaining the fire at night, and the size of the communications required for effective ventilation, must prevent its being so extensively useful for such purposes as might be imagined. In all cases where I have endeavored to apply its action, the objections made appeared to me sufficiently cogent to force me to give up the idea of trusting to it alone, as the effective means of inducing a sufficient amount of ventilation; but, though not resorted to for this purpose, its power should not be thrown away. By establishing a communication between it and the hold, or any part of the ship where ventilation may be more urgently required, a certain benefit may be always secured, which will be productive of much good, even where nothing else is attempted.

" The freedom of access generally necessary around the fire, and the construction of the cooling apparatus, does not afford facilities for those large and air-tight channels which are so essential for obtaining considerable power from any heating apparatus; the larger the area, however, and the more direct their access to the fire, or to any channel formed between hot plates, the greater is the discharging power. It is almost superfluous to observe, that all such air-channels should be carefully protected by several folds of wire gauze from the ingress of any spark, as without this there would not be a sufficient guard against the risk of accident by fire. The galley-fire operates most effectually when it is supplied with air by air-tight channels, solely from the places to be ventilated; but this is incompatible with its use as an open fireplace.

" In the temporary application of local fires on board ship, for the purpose of drying or effecting a change of air, much may often be done to render them more useful by studying the circumstances under which they are applied, and taking care that a distinct course shall be given for the ingress of fresh air to supply the fuel, and another for the discharge of the vitiated air or products of combustion before they have cooled so far as to lose their ascending power. When this is not attended to, the cold air supplying the fuel descending through the same channel as that through which the escape takes place, the conflicting currents retard each other's progress; the vitiated air gets too cold, and often remains a considerable time, to the danger or annoyance of those who are near it."

Ventilation of Ships by an Injector.

Even without a steam-engine and blower, a small boiler with a Körting's injector, to exhaust the air from the bottom of a ship daily, is a practical and satisfactory apparatus. It may be depended on if duly proportioned to the work to be done, and regularly used. The carrying of such a boiler would be a very simple matter in any ship, and require no engineer. The tubular connections with the interior of the ship should be directed to the lower levels, where the air is most commonly foul, and to all those parts where it may be at any time stagnant. It must not be forgotten that yellow fever has very frequently broken out when the hold of a ship has been opened in port after a passage across the Atlantic without disease manifestations. There is no reason whatever not to guard against the possibility of any such accidents, for forced ventilation will prevent them.

In planning for the systematic ventilation of ships, attention must be paid :

1. To the effectual removal or exhaust of foul air from the lowest parts by positive agency.

2. Admitting air through ample apertures, under such conditions that by the suction from below an equal distribution throughout the ship may occur.

3. Constructing channels, by partitions or otherwise, to give direction to the air-currents.

4. Means of cooling and drying the air, especially in the tropics.

Prevention of Yellow Fever in Cities.

My observations on this subject shall be few. My object is to direct attention to the fact that cities must remain in close maritime intercourse to suffer from the scourge. The many measures required to insure a generally healthy state of seaport towns do not admit of discussion here. There is one, however, which bears on my proposal to resort to cold and free ventilation, as the fundamental means on which we must depend in controlling yellow fever. It relates to the ventilation of cities and of residences.

It has often been remarked that the great fire of London annihilated the plague and established a solid groundwork for the city which has since flourished. Fire is a great destroyer of contagion and unwholesome dwellings. If extensive enough in a town, it may, as it not infrequently does, lay the foundation for great social and sanitary improvements. Chicago is a recent instance. The cheapest thing for the country that perhaps might have happened last year would have been a conflagration calculated to put Memphis in ashes. I am informed that a forest fire around Vera Cruz, were such a thing possible, might save it if the cleared area were afterward kept clear by human industry. Such radical purifications are accidental, and the sanitarian has to fight inch by inch, and step by step, against the habits of a people, the vested interests of property-holders, and the intrigues, cupidity, and ignorance of local authorities. It is an example of the highest form of statesmanlike care for the public health that we can adduce in the bold remodeling of so solid a city as Birmingham, England. Equally wonderful, though springing from different circumstances, is the provision made in many northern cities of the United States for health, comfort, and the luxurious enjoyment of life. The South undoubtedly demands unremitting attention to town sanitation. So do the vineyards of France demand good culture, but no amount of vigilance would arrest the phylloxera, in the absence of such definite measures as experience suggests to strike at the very root of the disease. Local measures are therefore, sometimes, secondary to measures from a distance, and this in my opinion will prove to be the case with yellow fever. Clear the shipping of the disease, and the towns will almost take care of themselves.

The recent epidemics have led to a natural awakening on hygienic matters. They have brought the subject of disease-prevention to the front rank among social and political questions, and the common enemy of the people of the United States is a preventible pestilence. Medical men of the highest qualifications have been marshaled as a Board to direct and encourage local measures. To doubt the beneficent result of such changes and reforms is to doubt the advantages of acquiring knowledge, or of that lever without which science is valueless, viz., method in its application.

Cold and Ventilation.

Cities liable to yellow fever can and should have the advantage of unlimited supply of cold ; or in other words, the evil effects of heat and excessive humidity should be counteracted, by the abstraction of the first and condensation of the last. Many efforts have been suggested, but few made, in this direction ; and since the careful study of the means for the refrigeration and ventilation of ships, I have pursued inquiries into the practical and economical means which can be employed for supplying houses with pure cold air, just as we supply houses with pure water. Impossible as the project has been deemed

hitherto, the saving I have been able to effect in operating freez-
ing-machines enables me to say that the great reform in tropical
habitations, which will surround the dwellers with a pure atmosphere, at any desired temperature and degree of humidity, is
attainable at moderate cost. My work is not completed, but, so
far as I have gone, it results from my calculations that every
house, in a sufficiently large group to warrant the investment,
can be supplied, at a cost which will compare favorably with
the cost of heating a New York residence, with pure filtered
air from above in a constant stream, driving out instead of
attracting from below the dangerous atmosphere of streets and
sewers, in such a city as New Orleans. Fifty cubic feet of air
per individual, at the typical temperature of 70° Fahr. and 70
humidity, can be flowing in constantly, rendering life in Havana, Vera Cruz, or Rio something very different from what it
is in the present sweating-houses—centers of human misery and
bodily enfeeblement. The details and calculations relating to
this proposal will form the subject of a special publication,
which is far advanced toward completion.

International Coöperation.

The ravages of yellow fever are, in a sense, altogether dispro-
portionate to the ruinous commercial sacrifices which it inflicts
on the Atlantic seaboard. The loss of life this year will be
trifling compared to the mortality by typhoid or genuine indigenous malarial fevers. But the maritime intercourse, and the
development of that immense ocean freight business which the
Mississippi is to feed from the West, must be dangerously crip-
pled by the inevitable quarantine regulations, which should give
way to the sanitary regulations and classification of shipping, as
elsewhere indicated.

The vigilance of the National Board of Health has brought
out very strikingly the benefits to be derived from watching for
infected shipping everywhere. In these days of Red-Cross Societies and International Commissions—of Plimsoll's Acts for
the protection of the seaman's life against the cupidity and
recklessness of ship-owners—it can not be difficult to carry the
science of disease-prevention into every seaport on the Atlantic
coast of America, and on to every ship floating in the yellow-

13

fever area. A commission of leading sanitary authorities, engineers, ship-builders, and ship-owners, nominated by the United States Government, the Mexican Republic, the Empire of Brazil, and the European governments whose colonies stud the western waters of the Atlantic, or whose navies skim the ocean, would in due time establish an understanding which would tend immensely to cement good-fellowship and encourage commerce. A preliminary and purely medical commission might investigate the soundness of that view, the acceptance of which, more than all else, will facilitate future operations, viz.: that yellow fever is, in its origin, simply and purely a nautical disease. It is the ships which in the first place have to be purified and reconstructed. It is said that a house never can be clean with a dirty kitchen: a seaport never can be pure with infected shipping.

CHAPTER VI.

The facts and observations recorded in the foregoing pages warrant us in stating, with considerable precision, some of the fundamental truths which are, perhaps for the first time, made evident and mutually dependent.

1. Yellow fever exists permanently in some of the ships sailing in the West Indian seas, extending eastward in the Atlantic probably as far as the 26th or 27th degree of west longitude, southward to the immediate vicinity of the equator, and northward to the calm-belt of Cancer. This is its *permanent* home.

2. The ships *obviously* or *latently* infected, by sojourn in these ocean limits, contaminate the ocean harbors chiefly and often, but less often the seaport towns themselves. These harbors and seaports cherish and retain the infection in direct relation to prevalent heat and moisture, hence to latitude, impurity by fecal or other contaminations, density of population, activity of trade, and proximity of the shore dwellings to the shipping. The number, age, and character of ships in harbor, the source and cleanliness or impurity of the said ships, the character of trade or purpose for which the shipping is used—whether in West Indian or colonial produce, liable to prompt decomposition, or ores, cotton, and other merchandise less likely to change —materially influence the severity, extent, and duration of an outbreak. The nature of the population afloat or in the harbor, in relation to susceptibility, is a potent factor under this head. Seaports and harbors constitute the *subpermanent* home of yellow fever.

3. River harbors and cities on river banks, above the points
of direct communication with the sea, obeying the same conditions as the above, constitute the occasional *transient* homes of
yellow fever.

4. It *must* be accepted as conclusively proved, that yellow
fever never has existed *permanently*, as cholera has done in Bengal, in any region—island or continent—*on land*. Yellow fever
is simply accidental, and promptly extinguished the farther from
shore or the higher the ascent up hills and mountains. Refugees
from infected shore or ocean centers locate the *rare and inhospitable* homes of yellow fever. Therefore, in the interior of spa-
cious islands and on the mighty continent of America, it never
did, never could, and never can live—much less originate.

5. Yellow fever is not one of the pure contagia, propagated
or reproduced, through time and space, by a specific virus or
diseased animal secretion, or definite migrating parasite, capable of communication only from the sick to the healthy, in any
latitude.

The "germ theory," as popularly understood, and the dependence of yellow fever on "specific spores" and "microphytes,"
may be regarded as having led to *fallacious* reasoning. There
is no foundation in *fact* for the plausible explanations of the
propagation of the disease by the reproduction of the lowest organisms.

6. Yellow fever is not personally contagious, but transmitted
by definite, or "particulate," putrescent emanations which foul
a confined atmosphere. Ill-ventilated rooms, especially below
the level of streets, deep ships with close and confined air-
space, are the favored sites of greatest malignancy. The products of a specific naval or marine decomposition adhere to
clothes, woolen goods, sails, and other articles transported by
man, and, with heat, moisture, and stagnant atmosphere, reproduce the poison, which the vital tissues and vital organs of
the stoutest and healthiest men can not resist.

7. Yellow fever is due to that form or specific kind of decomposition which demands the sea-water of the defined area
named, with, possibly, the ocean atmosphere charged with the
organic products in this region (of high humidity and constant
temperature of 80° or 90° Fahr.), besides sundry organic, ligne-

ous, or hydro-carbonaceous substances capable of fermentation. The susceptible crew or passengers, confined (especially in storms) in a space totally unequal to the wants of respiration, when ports and hatches are closed, discharging, as all human beings do, much moisture and effete animal matter, contribute to the pestilential cause, and are almost victims of self-poisoning—unwitting and unwilling suicides and plague-propagators.

8. The nature of yellow fever is best expressed by naming it Equatorial Atlantic Ocean typhus. *Fièvre matelote*, or sailor's fever, and nautical typhus, are the next best terms ; but there are other fevers of the sea, such as pure typhus, even in northern waters, from which it must be distinguished. It is the analogue of the old jail fever of the British prisons, but has never been known to acquire its distinctive characters on land, except as an ocean pestilence.

9. The distinctive signs of yellow fever are those of a putrid disease in which assimilation is suspended and the tissues degenerate. There is no tendency to high organic development in its course, but only to retrograde tissue-changes and the dissolution of organs, explaining the incurability of the disease. The *ante-mortem* decomposition is attended with arrest of secretions, and the diffusion and dispersion of their usual ingredients throughout the system, whereby it is also poisoned ; ejection of dead and disintegrated blood and tissues ; progressively reduced functional activity, typified by the slow pulse ; active chemical changes in the devitalized tissues without equalizing heat influences—hence the continued high temperature ; lowered or lost sensibility and profound cerebral changes—hence the occasional delirium and the *mieux de la mort*, or " calm before death." Fatality or malignancy seems closely related to the quantity of putrescent matter absorbed by the susceptible individual in close quarters— hence the terrible consequences of ship infection.

10. The phenomena after death are those of rapid decay, a continuous decomposition, sometimes with actual *post mortem* increase of temperature, in part analogous to the changes in actual life.

11. Treatment, in medical language, can only be expectant and palliative. No specific cure ever has been or can be found.

A case of virulent infection implies death as certainly as a dagger-thrust into the heart. By the most judicious nursing (the retrograde tissue-changes ceasing) recovery occurs naturally. The obvious antidotes are antiseptics, which, unfortunately, can ill operate on the dying molecular matter of a still living man. Cold usually accelerates death.

12. Nature's antidotes against putrefaction are dryness, heat, and cold. The effect of frost, in a yellow-fever city, is more marked, because more prompt, than the effect of cold on the decomposition of ordinary animal and vegetable products. The sudden arrest of the disease, even in the homes of a suddenly cooled city, shows that the products of specific yellow-fever decomposition can only be produced and maintained at high temperatures. The refrigeration of ships will purify them, but all foci of after-contamination should be removed—all chance of renewed and regenerated infection guarded against. The same reason that accounts for butchers' meat not decomposing in winter, as it does in summer, explains how cold arrests yellow fever. There is a difference in degree in favor of the arrest of that specific decomposition, to which I have alluded in relation to this disease. When heat is used, the well-known conditions essential to the destruction of putrid elements in organic liquids, as established by Pasteur and Wyman, must be attended to, viz., *prolonged* exposure to 212° Fahr. and upward.

13. Prevention demands sound ships and pure dry air in abundance. Every sailor should have, and can have, from 1,000 to 2,000 cubic feet per hour, provided adequate appliances are compulsorily introduced into the marine. Clean ships, and the absolute exclusion from the area of probable ocean infection of old and rotten vessels, must be secured, *coute qui coute;* and the combined navies of the world, deeply interested in the extinction of this plague of the sea, may and should join hand in hand to facilitate the inspection and overhauling of all perilous and infected craft.

14. The concealment of ship infection, the shipping of a healthy seaman or a healthy crew on an infected ship, and the landing of people and things from ships without adequate disinfection, should be made penal offenses by the law of maritime nations; and the penalties against all offenders should be severe,

such as total loss of certificate for officers, with imprisonment without option of a fine.

These are by no means all the conclusions I might draw, but they are those of prime practical moment, calculated to disabuse the public mind as to the *inevitable* recurrence of this fearful malady. Strong in my convictions, earnest in my work, and positive as to the benign influence of a true enlightenment, I have tapped the precious source of that flood of light which must bear hope and comfort—a rational belief in future safety—wherever the fetid cause of yellow-fever mortality is desolating the human family.

INDEX OF AUTHORITIES.

GENERAL INDEX.

THE END.

MEDICAL WORKS,

PUBLISHED BY

D. APPLETON & COMPANY,

549 & 551 Broadway, New York.

PRICE

ANSTIE (FRANCIS E.) Neuralgia, and Diseases which resemble it. By Francis E. Anstie, M. D., F. R. C. P., Senior Assistant Physician to Westminster Hospital; Lecturer on Materia Medica in Westminster Hospital School; and Physician to the Belgrave Hospital for Children; editor of " The Practitioner " (London), etc. 1 vol., 12mo...Cloth, $2 50*

BARKER (FORDYCE). On Sea-Sickness. A Popular Treatise for Travelers and the General Reader. By Fordyce Barker, M. D., Clinical Professor of Midwifery and Diseases of Women in the Bellevue Hospital Medical College, etc. Small 12mo........Cloth, 75

—— On Puerperal Disease. Clinical Lectures delivered at Bellevue Hospital. A Course of Lectures valuable alike to the Student and the Practitioner. Third edition. In 1 vol., 8vo...........Cloth, 5 00*
Sheep, 6 00*

BARNES (ROBERT). Lectures on Obstetric Operations, including the Treatment of Hæmorrhage, and forming a Guide to the Management of Difficult Labor. By Robert Barnes, M. D., F. R. C. P., Fellow and late Examiner in Midwifery at the Royal College of Physicians; Examiner in Midwifery at the Royal College of Surgeons, London. 1 vol., 8vo. Third edition, revised and extended. Illustrated ....Cloth, 4 50*

BARTHOLOW'S Treatise on Materia Medica and Therapeutics. A new and revised edition. By Roberts Bartholow, M. A., M. D., Professor of Materia Medica and Therapeutics in the Jefferson Medical College; late Professor of the Theory and Practice of Medicine, and of Clinical Medicine, and formerly Professor of Materia Medica and Therapeutics, in the Medical College of Ohio. Revised edition of 1879. 1 vol., 8vo. 548 pages........ ......Cloth, $5.00; Sheep, 6 00*

—— **Practice of Medicine**......................(In press.)

BASTIAN (H. CHARLTON, M. D., F. R. S.) On Paralysis from Brain-Disease in its Common Forms. 1 vol., 12mo.............Cloth, 1 75

—— **Diseases of Nerves and Spinal Cord**...(In press.)

BENNET (J. H.) Winter and Spring on the Shores of the Mediterranean; or, The Riviera, Mentone, Italy, Corsica, Sicily, Algeria, Spain, and Biarritz, as Winter Climates. By J. Henry Bennet, M. D., Member of the Royal College of Physicians, London. With numerous Illustrations. 1 vol., 12mo. New revised edition. Cloth, 3 50*

—— On the Treatment of Pulmonary Consumption, by Hygiene, Climate, and Medicine. 1 vol., thin 8vo...................Cloth, 1 50*

BILLROTH (Dr. THEODOR). General Surgical Pathology and Therapeutics. A Text-book for Students and Physicians. By Dr. Theodor Billroth, Professor of Surgery in Vienna. From the eighth
German edition, by special permission of the author, by Charles E.
Hackley, M. D., Surgeon to the New York Eye and Ear Infirmary;
Physician to the New York Hospital. **Fourth American edition, revised
and enlarged.** 1 vol., 8vo..................Cloth, $5.00*; Sheep, $6 00*

BUCK (GURDON). Contributions to Reparative Surgery, showing its
Application to the Treatment of Deformities, produced by Destructive Disease or Injury; Congenital Defects from Arrest or Excess
of Development; and Cicatricial Contractions following Burns.
Illustrated by Thirty Cases and fine Engravings. 1 vol., 8vo. .Cloth,   3 00*

CARPENTER (W. B.) Principles of Mental Physiology, with their
Application to the Training and Discipline of the Mind, and the
Study of its Morbid Conditions. By William B. Carpenter, M. D.,
LL. D., F. R. S., F. L. S., F. G. S., Registrar of the University of
London, etc. 1 vol., 12mo.............................Cloth,   3 00

CHAUVEAU. The Comparative Anatomy of Domesticated Animals.
By A. Chauveau, Professor at the Lyons Veterinary School. Sec-
ond edition, revised and enlarged, with the Coöperation of S.
Arlong, Professor at the Toulouse Veterinary School. Translated
and edited by George Fleming, F. R. G. S., M. A. I., etc. With 450
Illustrations. 1 vol., 8vo...............................Cloth,   6 00

COMBE (ANDREW). The Management of Infancy, Physiological and
Moral. By Andrew Combe, M. D. Revised and edited by Sir James
Clark, Bart., K. C. B., M. D., F. R. S. 1 vol., 12mo........Cloth,   1 50

DAVIS (HENRY G.) Conservative Surgery. With Illustrations. 1
vol., 8vo.. ... .Cloth,   3 00*

ECKER (ALEXANDER). Convolutions of the Brain. Translated
from the German by Robert T. Edes, M. D. 1 vol., 8vo.....Cloth,   1 25*

ELLIOT (GEORGE T.) Obstetric Clinic: A Practical Contribution to
the Study of Obstetrics, and the Diseases of Women and Children.
By George T. Elliot, Jr., A. M., M. D. 1 vol., 8vo.........Cloth,   4 50*

FLINT'S Manual of Chemical Examinations of the Urine in Disease;
with Brief Directions for the Examination of the most Common
Varieties of Urinary Calculi. By Austin Flint, Jr., M. D. 1 vol.
Revised edition............. ....Cloth,   1 00*

—— Physiology of Man. Designed to represent the Existing State of
Physiological Science as applied to the Functions of the Human
Body. By Austin Flint, Jr., M. D., Professor of Physiology and
Microscopy in the Bellevue Hospital Medical College, New York;
Fellow of the New York Academy of Medicine; Member of the
Medical Society of the County of New York; Resident Member
of the Lyceum of Natural History in the City of New York, etc.
Complete in 5 vols.

Vol. 1. Introduction; The Blood; Circulation; Respiration. 8vo.
Vol. 2. Alimentation; Digestion; Absorption; Lymph, and
Chyle. 8vo.
Vol. 3. Secretion; Excretion; Ductless Glands; Nutrition; Ani-
mal Heat; Movements; Voice and Speech. 1 vol., 8vo.
Vol. 4. The Nervous System. 1 vol., 8vo.
Vol. 5. (With a General Index to the five volumes.) Special Senses;
Generation. Per vol...........Cloth, $4.50*; Sheep,   5 50*
The five volumes...............Cloth, $22.00*; Sheep, 27 00*

FLINT'S Text-Book of Human Physiology; designed for the Use of Practitioners and Students of Medicine. Illustrated by three Lithographic Plates, and three hundred and thirteen Woodcuts. *Second edition, revised.* 1 vol., imperial 8vo....... Cloth, $6.00* ; Sheep, $7 00*

—— The Physiological Effects of Severe and Protracted Muscular Exercise; with Special Reference to its Influence upon the Excretion of Nitrogen. By Austin Flint, Jr., M. D., etc. 1 vol., 12mo. Cloth, 1 00

—— **The Source of Muscular Power.** Arguments and Conclusions drawn from Observation upon the Human Subject under Conditions of Rest and of Muscular Exercise................ 1 00*

FREY (HEINRICH). The Histology and Histochemistry of Man. A Treatise on the Elements of Composition and Structure of the Human Body. By Heinrich Frey, Professor of Medicine in Zurich. Translated from the fourth German edition, by Arthur E. J. Barker, Surgeon to the City of Dublin Hospital; Demonstrator of Anatomy, Royal College of Surgeons, Ireland; Visiting Surgeon, Convalescent Home, Stillorgan; and revised by the author. With 608 Engravings on Wood. 1 vol., 8vo.....................Cloth, $5.00*; Sheep, 6 00*

HAMILTON (ALLAN McL., M. D.) Clinical Electro-Therapeutics, Medical and Surgical. A Hand-Book for Physicians in the Treatment of Nervous and other Diseases. 1 vol., 8vo.........Cloth, 2 00*

HAMMOND (W. A.) A Treatise on Diseases of the Nervous System. By William A. Hammond, M. D., Professor of Diseases of the Nervous System and of Clinical Medicine in the Bellevue Hospital Medical College; Physician-in-Chief to the New York State Hospital for Diseases of the Nervous System, etc. New edition, with 109 Illustrations. Rewritten, enlarged, and improved. 1 vol., large 8vo.
Cloth, $6.00*; Sheep, 7 00*

—— Clinical Lectures on Diseases of the Nervous System. Delivered at Bellevue Hospital Medical College. Edited by T. M. B. Cross, M. D. 1 vol., 8vo...Cloth, 3 50*

HEALTH PRIMERS. Edited by J. Langdon Down, M. D., F. R. C. P.; Henry Power, M. B., F. R. C. S.; J. Mortimer-Granville, M. D., and John Tweedy, F. R. C. S. Square 16mo..............Cloth, each 40
Now Ready:

I. Exercise and Training.	IV. The House and its Surroundings.
II. Alcohol: Its Use and Abuse.	V. Personal Appearance in Health
III. Premature Death: Its Promotion or Prevention.	and Disease.
	VI. Baths and Bathing.

HOFFMANN–ULTZMANN. Introduction to an Investigation of Urine, with Special Reference to Diseases of the Urinary Apparatus. By M. B. Hoffmann, Professor in the University of Gratz, and R. Ultzmann, Tutor in the University of Vienna. Second enlarged and improved edition.... ....(*In press.*)

HOFFMANN (FREDERICK, Ph. D., Pharmaceutist in New York). Manual of Chemical Analysis, as applied to the Examination of Medicinal Chemicals. A Guide for the Determination of their Identity and Quality, and for the Detection of Impurities and Adulterations. For the Use of Pharmaceutists, Physicians, Druggists, and Manufacturing Chemists and Students....................Cloth, 3 00*

HOLLAND (Sir HENRY). Recollections of Past Life. Reminiscences of Men, Manners, and Things. 1 vol., 12mo..............Cloth, 2 00

4

PRICE

NEW YORK MEDICAL JOURNAL. Edited by James B. Hunter, M. D.
The largest medical monthly published.........Terms per annum, $4 00*
Specimen numbers sent by mail on receipt of 25 cents.

NIEMEYER (Dr. FELIX VON). A Text-Book of Practical Medicine,
with Particular Reference to Physiology and Pathological Anatomy.
Containing all the author's Additions and Revisions in the eighth
and last German edition. Translated from the German edition, by
George H. Humphreys, M. D., and Charles E. Hackley, M. D. 2
vols., 8vo...............................Cloth, $9.00*; Sheep, $11 00*

NIGHTINGALE'S (FLORENCE) Notes on Nursing. What it is, and
what it is not. 1 vol., 12mo..........................Cloth, 75

PAGET. Clinical Lectures and Essays. By Sir James Paget, Bart.,
F. R. S., D. C. L., Oxon., LL. D., Cantab, etc. Edited by Howard
Marsh, F. R. C. S., etc. 1 vol., 8vo......................Cloth, 5 00*

PEASLEE (E. R.) A Treatise on Ovarian Tumors; their Pathology,
Diagnosis, and Treatment, with Reference especially to Ovariotomy.
By E. R. Peaslee, M. D., LL. D., Professor of Diseases of Women,
in Dartmouth College; one of the Consulting Surgeons to the New
York State Woman's Hospital; formerly Professor of Obstetrics
and Diseases of Women, in the New York Medical College; Corresponding Member of the Obstetrical Society of Berlin, etc. In one
large vol., 8vo. With Illustrations.........Cloth, $5.00*; Sheep, 6 00*

PEREIRA'S (Dr.) Elements of Materia Medica and Therapeutics.
Abridged and adapted for the Use of Medical and Pharmaceutical
Practitioners and Students, and comprising all the Medicines of the
British Pharmacopœia, with such others as are frequently ordered
in Prescriptions, or required by the Physician. Edited by Robert
Bentley and Theophilus Redwood. New edition. 1 vol., royal 8vo.
Cloth, $7.00*; Sheep, 8 00*

REPORTS. Bellevue and Charity Hospital Reports for 1870. Containing
Valuable Contributions from Isaac E. Taylor, M. D., Austin Flint,
M. D., Lewis A. Sayre, M. D., W. A. Hammond, M. D., T. Gaillard
Thomas, M. D., Frank H. Hamilton, M. D., and others. 1 vol., 8vo.
Cloth, 4 00*

RICHARDSON. Diseases of Modern Life. By Benjamin Ward Richardson, M. D., M. A., F. R. S., Fellow of the Royal College of Physicians. 1 vol., 12mo.............. ....Cloth, 2 00

ROSCOE and SCHORLEMMER. A Treatise on Chemistry by H. E.
Roscoe, F. R. S., and C. Schorlemmer, F. R. S., Professors of Chem-
istry in Owens College, Manchester. To be completed in three vol-
umes. Fully illustrated.
Now Ready:
Vol. I. The Non-metallic Elements. 8vo....... .....Cloth, 5 00
Vol. II., Part I. Metals. 8vo.................. ....Cloth, 3 00
In Preparation:
Vol. II., Part II. Metals.

SAYRE (LEWIS A., M. D.) Practical Manual of the Treatment of
Club-Foot. By Lewis A. Sayre, M. D., Professor of Orthopedic
Surgery in the Bellevue Hospital Medical College; Surgeon to
Bellevue Hospital, etc. New edition. 1 vol., 12mo........Cloth, 1 25*

—— Lectures on Orthopedic Surgery and Diseases of the Joints, delivered at Bellevue Hospital Medical College during the Winter Session of 1874-'75, by Lewis A. Sayre, M. D ...Cloth, $5.00*; Sheep, 6 00*

SCHROEDER (Dr. KARL). A Manual of Midwifery, including the
Pathology of Pregnancy and the Puerperal State. By Dr. Karl
Schroeder, Professor of Midwifery, and Director of the Lying-in In-
stitution, in the University of Erlangen. Translated into English from
the third German edition, by Charles H. Carter, B. A., M. D., B. S.,
London; Member of the Royal College of Physicians, London, etc.
With 26 Engravings on Wood. 1 vol., 8vo...Cloth, $3.50*; Sheep, $4 50*

SIMPSON (Sir JAMES Y.) Selected Obstetrical and Gynæcological
Works of Sir James Y. Simpson, Bart., M. D., and late Professor
of Midwifery in the University of Edinburgh. Edited by J. Watt
Black, M. A., M. D., M. R. C. P. L.; Physician Accoucheur to Char-
ing Cross Hospital, London, and Lecturer on Midwifery and the
Diseases of Women and Children in the Hospital School of Medicine. 1 vol., 8vo.................Cloth, $3.00*; Sheep, 4 00*

——— Anæsthesia, Hospitalism, etc. By Sir James Y. Simpson, Bart.,
M. D. Edited by Sir Walter Simpson, Bart..Cloth, $3.00*; Sheep, 4 00*

SIMPSON (Sir JAMES Y.) The Diseases of Women. By Sir James Y.
Simpson, Bart., M. D. Edited by Alexander Simpson, M. D., Pro-
fessor of Midwifery in the University of Edinburgh.
Cloth, $3.00*; Sheep, 4 00

SMITH (EDWARD, M. D., LL. B., F. R. S.) Foods. (*International Scientific Series.*)............................ .... 1 75

——— Health: A Hand-Book for Households and Schools. 12mo..Cloth, 1 00

STEINER. Compendium of Children's Diseases: A Hand-Book for
Practitioners and Students. By Dr. Johannes Steiner, Professor of
Diseases of Children in the University of Prague, etc. Translated
from the second German edition, by Lamson Tait, F. R. C. S., etc.
1 vol., 8vo.............................Cloth, $3.50*; Sheep, 4 50*

STROUD (WILLIAM, M. D.) The Physical Cause of the Death of
Christ, and its Relations to the Principles and Practice of Christian-
ity. With Letter on the Subject by Sir James Y. Simpson, Bart.,
M. D. 12mo...Cloth, 2 00

SWETT. A Treatise on the Diseases of the Chest. Being a Course of
Lectures delivered at the New York Hospital. By John A. Swett,
M. D. 1 vol., 8vo..............................Cloth, 3 50*

TILT'S Hand-Book of Uterine Therapeutics. Second American edition,
revised and amended. 1 vol., 8vo, 368 pages...... ....Cloth, 3 50*

VAN BUREN (W. H.) Lectures upon Diseases of the Rectum, delivered at the Bellevue Hospital Medical College, Session 1869–1870,
by W. H. Van Buren, A. M., M. D., Professor of the Principles of
Surgery, with Diseases of the Genito-Urinary Organs, etc., in the
Bellevue Hospital Medical College; one of the Consulting Surgeons
of the New York Hospital, of the Bellevue Hospital; Member of
the New York Academy of Medicine, of the Pathological Society
of New York, etc. 1 vol., 12mo............. ....Cloth, 1 50*

——— A Practical Treatise on the Surgical Diseases of the Genito-Urinary Organs, including Syphilis. Designed as a Manual for Students
and Practitioners. With Engravings and Cases. By W. H. Van
Buren, A. M., M. D., and Edward L. Keyes, A. M., M. D., Professor
of Dermatology in Bellevue Hospital Medical College, Surgeon to
the Charity Hospital, Venereal Division; Consulting Dermatologist
to the Bureau of Out-Door Relief, Bellevue Hospital, etc. 1 vol.,
8vo..Cloth, $5.00*; Sheep, 6 00*

PRICE

VOGEL (A.) A Practical Treatise on the Diseases of Children. By Alfred Vogel, M. D., Professor of Clinical Medicine in the University of Dorpat, Russia. Translated and edited by H. Raphael, M. D., late House Surgeon to Bellevue Hospital, Attending Physician for the Eastern Dispensary for the Diseases of Children, etc. From the fourth German edition. Illustrated by Six Lithographic Plates. 1 vol., 8vo............... ....Cloth, $4.50*; Sheep, $5 50*

WAGNER (RUDOLF). Hand-Book of Chemical Technology. Translated and edited from the eighth German edition, with extensive Additions, by William Crookes, F. R. S. With 336 Illustrations. 1 vol., 8vo. 761 pages......................... ....Cloth, 5 00*

WALTON (GEORGE E., M. D.) Mineral Springs of the United States and Canadas. Containing the latest Analyses, with full Description of Localities, Routes, etc. 12mo.................... Cloth, 2 00

WELLS (Dr. T. SPENCER). Diseases of the Ovaries. 1 vol., 8vo... 4 50*

WYLIE (W. GILL, M. D.) Hospitals: History of their Origin, Development, and Progress, during the First Century of the American Republic. Boylston Prize-Essay of Harvard University for 1876. 1 vol., 8vo.. ......Cloth, 2 50*

Appletons' Journal.
PUBLISHED MONTHLY.

The proprietors of APPLETONS' JOURNAL will henceforth devote it exclusively to literature of a high order of excellence, by writers of acknowledged standing.

It is the growing habit of the leading minds in all countries to contribute their best intellectual work to the magazines and reviews ; and, in order that APPLETONS' JOURNAL may adequately reflect the intellectual activity of the time thus expressed, it will admit to its pages a selection of the more noteworthy, critical, speculative, and progressive papers that come from the pens of these writers.

Fiction will still occupy a place in the JOURNAL, and descriptive papers will appear; but large place will be given to articles bearing upon literary and art topics, to discussions of social and political progress, to papers addressed distinctly to the intellectual tastes of the public, or devoted to subjects in which the public welfare or public culture is concerned.

Subscription, $3.00 per annum; single copy, 25 cents, post-paid. Binding-cases, 50 cents each. Specimen copy, 18 cents.

The Popular Science Monthly.
Conducted by E. L. and W. J. YOUMANS.

With the number for January, 1879, THE POPULAR SCIENCE MONTHLY was permanently enlarged to 144 pages. Its readers will thus receive sixteen pages additional matter without increase of price, and the editors will, at the same time, be enabled to make it a more complete exponent of current scientific thought. The contents of the magazine will, as heretofore, consist of original scientific articles from eminent home and foreign writers, selections, falling within its scope, from the leading English periodicals, translations from foreign languages, synopses of important scientific papers, and notes of the progress of science throughout the world.

THE POPULAR SCIENCE MONTHLY is a large octavo, handsomely printed on clear type, and, when necessary to further convey the ideas of the writer, fully illustrated.

TERMS : $5.00 per annum, or 50 cents per number. Postage prepaid. Binding-cases for any volume will be forwarded by mail, post-paid, upon receipt of 50 cents. Specimen copy, 35 cents.

The North American Review.
PUBLISHED MONTHLY.

This old and valued periodical, under new and energetic management, has during the past year stepped into the front rank of literature, showing itself the equal, if not the superior, of the great Reviews and Quarterlies of the Old World. Per annum, $5.00 ; per number, 50 cents.

D. APPLETON & CO., PUBLISHERS.

TO THE MEDICAL PROFESSION.

WE beg to call your attention to the merits of the NEW YORK MEDICAL JOURNAL. In doing so, we can say with confidence that this journal occupies a higher place in medical literature than any other monthly publication in this country, and that henceforth no effort will be spared to enhance its value, and render it indispensable to every practitioner who desires to keep up with the times. Trusting you will favor us with your support,

We are, yours truly,

D. APPLETON & CO.

The foremost American Monthly.

THE

NEW YORK MEDICAL JOURNAL,

Edited by JAMES B. HUNTER, M. D.,

Surgeon to the New York State Woman's Hospital; Consulting Surgeon to the New York Infirmary for Women and Children; Member of the New York Obstetrical Society, etc.

The leading features of this Journal are the following:

ORIGINAL COMMUNICATIONS FROM EMINENT MEMBERS OF THE PROFESSION.
REPORTS OF INTERESTING CASES IN PRIVATE PRACTICE.
NOTES OF PRACTICE IN METROPOLITAN HOSPITALS, ILLUSTRATING THE USE OF NEW METHODS AND NEW REMEDIES.
TRANSLATIONS AND EXTRACTS GIVING THE CREAM OF ALL THE FOREIGN JOURNALS.
REPORTS ON MEDICINE, SURGERY, OBSTETRICS, GYNÆCOLOGY, LARYNGOLOGY, PATHOLOGY, etc.
CRITICAL AND IMPARTIAL REVIEWS OF ALL NEW MEDICAL BOOKS.
PROCEEDINGS OF MEDICAL SOCIETIES.
COPIOUS ILLUSTRATIONS BY MEANS OF WOODCUTS.
THE LATEST GENERAL MEDICAL INTELLIGENCE.

A new volume of the NEW YORK MEDICAL JOURNAL begins with the numbers for January and July each year. Subscriptions received for any period.

Terms, $4.00 per Annum, postage prepaid by the Publishers.

Trial Subscriptions will be received at the following rates: Three months, $1.00; six months, $2.00; specimen copy, 25 cents.

A General Index to the NEW YORK MEDICAL JOURNAL, from its first issue to June, 1876—including twenty-three volumes—now ready. Price, in cloth, 75 cents, post-paid. *Remittances, invariably in advance, should be made to*

D. APPLETON & CO., Publishers, 549 & 551 Broadway, N. Y.